GORY DETAILS

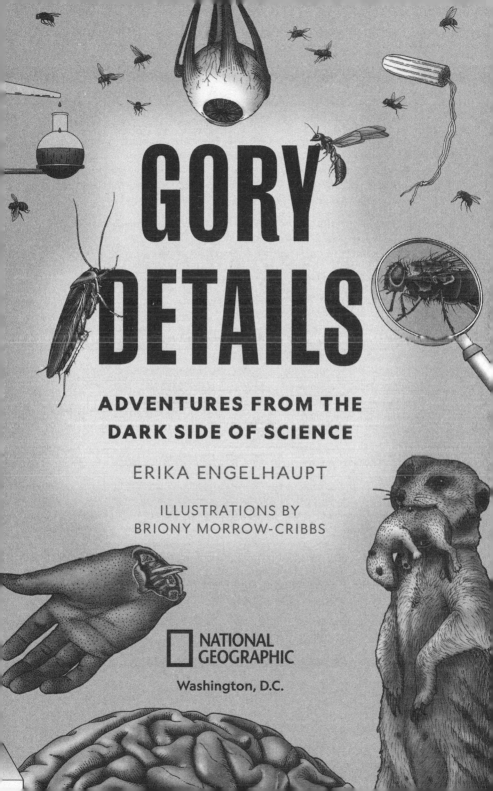

GORY DETAILS

ADVENTURES FROM THE DARK SIDE OF SCIENCE

ERIKA ENGELHAUPT

ILLUSTRATIONS BY
BRIONY MORROW-CRIBBS

NATIONAL GEOGRAPHIC

Washington, D.C.

Published by National Geographic Partners, LLC
1145 17th Street NW Washington, DC 20036

Library of Congress Cataloging-in-Publication Data

Names: Engelhaupt, Erika, author.
Title: Gory details : adventures from the dark side of science / Erika
    Engelhaupt.
Description: Washington, DC : National Geographic Partners, LLC, [2020] |
    Includes bibliographical references and index.
Identifiers: LCCN 2019044364 (print) | LCCN 2019044365 (ebook) | ISBN
    9781426220975 (hardback) | ISBN 9781426220982 (ebook)
Subjects: LCSH: Science--Miscellanea. | Medicine--Miscellanea. |
    Curiosities and wonders.
Classification: LCC Q173 .E536 2020 (print) | LCC Q173 (ebook) | DDC
    500--dc23
LC record available at https://lccn.loc.gov/2019044364
LC ebook record available at https://lccn.loc.gov/2019044365

Since 1888, the National Geographic Society has funded more than 13,000 research, exploration, and preservation projects around the world. National Geographic Partners distributes a portion of the funds it receives from your purchase to National Geographic Society to support programs including the conservation of animals and their habitats.

Get closer to National Geographic explorers and photographers, and connect with our global community. Join us today at nationalgeographic.com/join

For rights or permissions inquiries, please contact National Geographic Books Subsidiary Rights: bookrights@natgeo.com

Interior design: Melissa Farris and Nicole Miller

Printed in the United States of America

20/MP-PCML/1

For Gay, Darell,
and Jay

CONTENTS

INTRODUCTION

Until I was seven years old, my family lived among the rolling hills on the outskirts of Kansas City, Missouri. Our little white stucco house was perched at the top of a gentle peak, tailed by a long, curving driveway. Each afternoon, the school bus would drop me at the bottom, where my mother and our big black German shepherd would be waiting for me.

One day, something new appeared at the end of the driveway: two mountains. (Well, hills, really, but keep in mind that I was small.) They were made of trash, piled high where a truck had pulled over and tipped its unloved cargo onto a convenient patch of land. As my mother and I approached, the jumble sorted itself into recognizable objects. There were filing cabinets and cardboard boxes filled with papers. My mom picked up a dark sheet and held it up to the light; it was an x-ray of teeth. We realized that a dentist's office must have closed down and dumped its entire contents at the end of our driveway.

Although toys from the waiting room were scattered in each pile, they were less interesting to me than the jewelry box I unearthed, holding a silver necklace strung with tiny jade green birds. But then I came upon the best part: plaster casts of patients' teeth, both uppers and lowers. Soon I was setting aside all the gnarliest entries: teeth with chips, teeth that tilted out like broken fence posts, teeth with missing teeth—the uglier, the better.

My parents were irritated by the dump at our doorstep; eventually, an uncle with construction equipment would spread and cover it, creating our own mini-landfill. But because it was there, I got to keep some of my favorite finds. I'm sure I was the only girl in those parts of Missouri who had not only her own playhouse but also a playhouse with windowsills featuring rows of snaggleteeth. Occasionally, I rearranged them, finding new and more hideous combinations of uppers and lowers. On breezy summer evenings, I could lie on my cot with the windows open and feel comforted by the grinning incisors glowing in moonlight.

I suppose if my parents had been horrified by my collection—or later, by my abiding interest in bloody nature documentaries and Stephen King novels—I might have turned out a bit differently. Maybe instead of writing about gross stuff, I'd be an accountant, or one of those people who gets queasy at the sight of blood.

But that was not my fate. A couple years later, my family moved to a swampy seven acres in Florida. My father, an engineer, built his own cinder block electrochemistry laboratory next to our mobile home. Inside those block walls, he tried to introduce me to a few basic scientific principles. I didn't understand much but was amazed at his ability to stick a penny into a tank of liquid and pull it out the next day, covered in shiny nickel. Most important, I learned that it was entirely possible to figure out how things work—and to figure it out using science.

Fast-forward about 30 years. After a decade of research (where I had engaged in activities like tromping around in swamps to study carbon-containing compounds), I became a writer and editor at *Science News* magazine in Washington, D.C. When the opportunity to write a blog came along, I knew I wanted in. It took only a glance at my office bookshelf—filled with titles like *Blood Work*, *The Killer of Little Shepherds,* and *That's Disgusting*—to spark a concept. Although I hadn't considered myself particularly interested in the macabre, I realized that my morbid curiosity had been there all along. And so the Gory Details blog was born.

Since then, I've had years of adventures writing about subjects that require warning when people ask me what I'm working on. For a while, colleagues would pass me any article or scientific paper that dealt with pee or poop. (I'm told that traffic to my post about what happens to urine in a swimming pool spikes every summer, around the time pools open for the season.) Later, when I went to work as a science editor at National Geographic's website, Gory Details came with me, and has lived there ever since.

Along the way, some of my favorite stories have ended up being the ones that I initially worried were too disturbing for prime time. For instance, when a colleague from *Science News* innocently asked whether it's true that pets sometimes eat their deceased owners, I thought I'd look into it in the spirit of shared curiosity. I was skeptical, though, about a story with such disturbing implications for animal lovers.

As it turned out, not only did a lot of other people have the same question, but there were also plenty of forensic case studies describing such incidents. Other reporters had already written about some of these, but I decided to apply my research skills to taking a deeper dive into the forensic journals. Despite my fears that dog lovers everywhere would turn on me, the piece ended up being one of the most popular articles on the National Geographic website that year. Seems that even if people deride a gross question, they still want to know the answer.

Beyond satisfying my own weird inquisitiveness, the larger goal of Gory Details has always been to create a place where it's OK to talk about gross, taboo, or morbid topics — and then to examine them, up close, through the lens of science.

Why would I want to spend my days thinking about topics that, on a good day, are unpleasant? What it comes down to is this: I'm less fearful of the things I've written about. When I look more closely at whatever rattles me—death, disease, creepy clowns— scientific analysis makes it a little more manageable.

Perhaps as a result of my own worst fears, death and murder have been go-to topics in my science reporting. I've delved into new forensic science techniques, as well as old-school methods of investigation, such as the miniature crime scenes used to train detectives and forensic examiners. Other times, the subjects I write about are not life ending, but rather life altering for those who experience them—as in the case of people suffering from delusions of infestation, whose lives have been overturned by their horrifying conviction that invisible bugs plague them.

That said, not everything I write about has such high stakes. It has always been my hope that Gory Details is as fun as it is informative. Sometimes, it's just a good place to ponder a gross question, like the worst possible insect sting or the surprisingly complex science of earwax.

In composing these essays, I look for topics with a push and a pull: something that at first makes me look away but that also keeps me peeking through the fingers clamped over my eyes. This is what writers like to call "dramatic tension"—and when I find dramatic tension alongside an intriguing bit of science, I, for one, am hooked.

This book brings together some of the most fascinating stories I've found reporting Gory Details over the years, which have been further expanded and updated in these pages. You'll also find new tales that I've had the pleasure of digging up just for you, dear reader. If you're up for a gory science snack, you can dip into any entry that strikes your fancy; it will make sense on its own. Or you can read a full chapter and make a meal of it.

In these pages, I've selected subjects with that push and pull: stories that make me want to know more. Each chapter is composed of related essays around a theme, from death ("Morbid Curiosity") to our deepest, darkest thoughts ("Mysterious Minds"). In all these areas, scientists are pushing the boundaries of our knowledge about the gross, frightening, and taboo, uncovering surprising truths about our minds, bodies, and world.

For me, exploring these subjects has been a reminder that we don't have to lose our childlike curiosity. I don't know what ever happened to my dental molds; they were probably tossed out in one move or another. But I still have the necklace with the jade green birds; it serves as a small reminder that treasure can hide in surprising places—places most people would never think to look. I take pleasure in finding beauty within the ugliness of the world, and order amid its chaos.

And it all starts with being willing to ask a question that might raise eyebrows. Sometimes these questions can be uncomfortable—but the answers are always intriguing. I hope you'll walk away from this book feeling a little bolder about asking weird questions. And I hope the answers will make you want to know more.

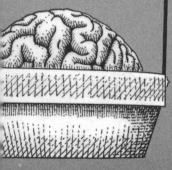

PART ONE

MORBID

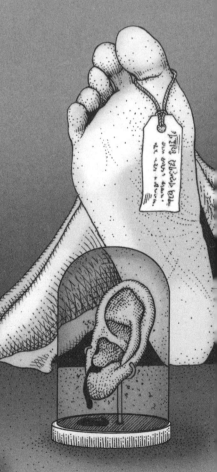

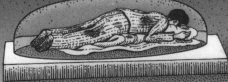

CURIOSITY

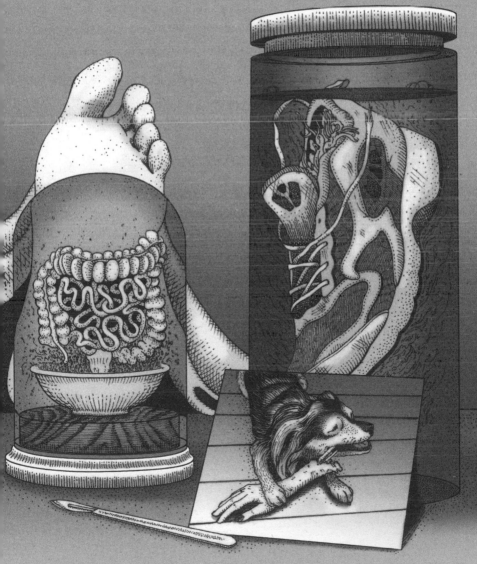

INTRODUCTION

NOT QUITE CSI

Morbid curiosity and the morgue

The first autopsy I ever witnessed was not quite what I had imagined. Growing up, I had always thought a postmortem exam would look like the grim dissections depicted on the medical drama *Quincy, M.E.* (though to be fair, not much more than a glimpse of a cadaver was allowed on early-1980s TV).

Ever since Jack Klugman portrayed Quincy, the crime-solving medical examiner, television has depicted autopsies in pretty much the same way: a basement morgue, often dimly lit except for bright lights hanging above a gurney; a wall of refrigerated drawers for bodies; and one of those hanging scales, into which the pathologist drops a wet, slippery heart or liver. There's also usually a viewing area, where family members look through a window as an assistant pulls a sheet back from their loved one's face.

The Quincy character—a cantankerous, brilliant, stubborn man who lived alone on a boat—helped establish the well-worn trope of the oddball medical examiner. And indeed, in most crime dramas since, there's generally been something a little weird about pathologists and forensic examiners: They talk to the bodies, and their matter-of-factness about mortality is off-putting (although that probably says more about the public than medical examiners; we're just not comfortable with people who are comfortable with death).

But some level of morbid curiosity is normal. We all fear the unknown—and death, as they say, is the great unknown. It's also the great certainty. We've devoted entire genres of art and literature to the subject, again and again seeking out the frisson that comes from brushing close to the grave and yet surviving. It's why we watch scary movies, and why we rubberneck at accidents, even though we complain that rubberneckers are ghouls as we drive away.

And although some decry the latest pop culture incarnations of murder tales as morbid entertainment, a preoccupation with death and violence is nothing new. These are the oldest stories in the book. The Bible is full of 'em, as are our most beloved tales, legends, myths, and fairy tales the world over.

What's new isn't the murder story, but the ways we tell it. Today, we can stream death—sometimes in graphic detail—into our homes 24/7, from true crime podcasts to the ID Channel. The enormous success of the genre is a testament to morbid curiosity's primordial pull.

Some evolutionary biologists explain this curiosity as a reasonable analysis of danger; we* look at death so that we can learn how to avoid it. Other animals exhibit this behavior too, including crows gathered around a deceased flock member while taking note of predators. Similarly, people fascinated by murder mysteries might analyze threats and glean potential avoidance strategies. (Recently I watched a television show called *I Survived* that consisted entirely of people describing how they got out of life-threatening situations; it was riveting.) The tagline of the hit true crime podcast *My Favorite Murder*—"Stay sexy, and don't get murdered"—also gives a nod to this idea.

Another explanation for morbid curiosity, as some psychologists have suggested, is that we're drawn to the macabre

----

* Though it's hard to determine experimentally *why* people have morbid curiosity, tests do show that the phenomenon is real. In 2017, Dutch psychologist Suzanne Oosterwijk reported on a series of experiments she had done to test for morbid curiosity. She found that people not only chose to see images depicting death or violence even if given the option not to, but they also would choose to get a longer look at these images over neutral or even pleasant ones.

because we actually yearn to empathize. We want to put ourselves in the unfortunate victim's shoes; it's part of our social nature. Others suggest we want to understand what creates a mind that yearns to harm others. If any of these theories hold water, it would seem that our intentions aren't evil.

My own morbid curiosity brought me to Baltimore's Office of the Chief Medical Examiner to attend a seminar on homicide investigation; this included the opportunity to observe autopsies. About the only thing that matched up with television were the scales. Oh, and one other small detail: a photograph of Jack Klugman, dressed in a white lab coat labeled ".J. Quincy," hung on the wall among pictures of former medical examiners.

I wanted to see what a real autopsy is like, and Baltimore— home to America's largest medical examiner's office—handles a lot of them. Inside the six-story Forensic Medical Center, 16 examiners perform about 4,000 autopsies a year on cadavers from across the state, accounting for about 10 percent of all deaths in Maryland. This includes not only homicides and accidents, but also just about anyone who's found dead unexpectedly. If you don't want to end up here, I'm told, die in the hospital, under the care of a doctor.

The Forensic Medical Center is a state-of-the-art facility completed in 2010, with "room to grow," says Bruce Goldfarb, assistant to the chief, as he whisks me through the enclosed ground-level garage. Gurneys roll straight into a freight elevator that takes them to the level above, where they can be rolled over a scale for a quick weigh-in, given a full-body x-ray* in 13 seconds, and then be

....

* The special x-ray machine that performs this function is made by Lodox Systems. Originally designed to prevent diamond-mine workers in South Africa from smuggling gems by swallowing or slipping them into an orifice, it features an ultralow-dose x-ray to quickly scan workers. Cadavers don't need to worry about their x-ray exposure, but the whole-body scan can help medical examiners spot bullets, tumors, and other injuries or illnesses.

wheeled into one of two autopsy rooms with eight bays. It's the shortest time you'll ever wait to see a doctor.

Then there are "decomp" cases, those bodies whose decomposition—read, odor—is too offensive for the general population of cadavers. These go to one of two smaller biohazard rooms that are specially ventilated, exchanging the air completely 30 times an hour. Still, nothing can eliminate the smell entirely; everyone leaving an occupied room carries with them a waft of pungent fatty acids that make up the signature smell of death.

Notably missing are the rows of refrigerated stainless steel drawers for cadavers that many people associate with morgues. Instead, there's just a chilled discharge room for bodies waiting to go to funeral homes after their autopsies. "There's no backlog," Goldfarb tells me. "The goal is for everyone to be in and out within 24 hours." You also won't find a viewing room with grieving family members identifying bodies. Here, the unknown are usually identified using photographs, dental records, fingerprints, or DNA.

On the first morning of my visit, there are 17 cadavers, including five in the decomposition rooms. By 8:30 a.m., the medical examiners are making their morning rounds. The divvying up is friendly and orderly; cases are divided equally among the five medical examiners, and no one argues over who has to do the decomps.

That day, five of the 17 dead arrived with gunshot wounds, noted as GSW on the list that's printed every morning and placed near the light blue booties everyone wears over their shoes. After donning mine, I walk into a high-ceilinged, brightly lit room. Cadavers are laid out on gurneys at seven of the eight autopsy stations.

The group moves briskly from body to body, nodding and noting details. In a couple of the gunshot wound cases, the police detectives assigned to the case are present and chime in with details: numbers of bullet holes observed, shell cartridges recovered, and eyewitness accounts.

After finishing both autopsy rooms, the forensic medicine student escorting me asks if I want to go in the decomposition room. The group is forging on, and I don't want to wimp out, so I agree. We enter a much smaller room, where three bodies lie in various hues of bruiselike yellow and purple. The first thing that hits you is the smell, which is bad, but not knock-you-over; no one wears a mask or plugs their nose.

When we leave, I'm feeling relieved until the student asks if I want to go to the other decomp room, too. The *other* decomp room? Once again, the group is headed that way, though the crowd's a bit thinner now. "In for a penny," I say, and we walk into the farthest room from the main autopsy theater. Here: two more bodies, both discovered in a more advanced state of decay. Within a few days after death, a body swells with the gases produced by bacteria, a state known as bloat. In this room I'm witnessing bloat and its accompanying processes of decomposition, including the odor.

That room sticks with me, literally, for the rest of the day. After an hour, I'm still smelling eau de decomposition, especially when I move, and I fear that the scent is wafting out of my long hair. "Don't worry, no one else can smell it on you," two students tell me, as they notice me sniffing my shirt in the ladies' room. I eye a can of deodorizing spray on the sink but decide against spraying myself in front of them. Later, I find the room empty, and I hiss a spray into the air and walk through the blessedly fragranced cloud. Even a shower that evening can't quite banish the smell, which by then may be stuck in my mind as much as in my nose.

The next morning, fresh from a second shower, I join the seminar students to observe the actual autopsy process from the relative comfort of the glass-walled viewing areas above the two main autopsy rooms. The procedure is a lot of work, with many details to attend to. Technicians and medical examiners collect various pieces of evidence—pants are bagged, bullets extracted, fingernails scraped.

In the other room, where there are no gunshot wounds, things are farther along. Six of the bodies have already been opened. Their chest cavities are splayed, yellow tendrils of fat hanging alongside red muscles and flaps of skin.

But your eyes go first to the heads. Several cadavers have their scalps peeled back, with skulls cut open and emptied of brains. The big empty holes are the most disturbing part of the scene. I had imagined that medical examiners would remove a cap at the top of the skull, but it's actually more like a quarter, the cut running across the top of the head to just above the ears, then across the back of the head.

An assistant is at work with the bone saw, and in just a few minutes, he removes a wedge of skull. Before you know it, he's reaching in to pull out the brain. A quick cut to the brain stem and it's free, pink and slicked with blood. He wraps it in a white towel and rolls it gently onto a plastic sheet on the hanging scale.

When you watch people who do this work every day, it's less shocking than it sounds. The medical examiners and assistants look calm and thoughtful. No one makes faces when something gross happens. It's just the reality of determining the cause and manner of death, which is a medical examiner's job. You have to examine every part of the body.

It's not pleasant, but I'm glad to have seen this. Should I die under mysterious circumstances, I feel more assured that my cadaver will be treated well and searched vigilantly for clues to the cause of my demise. The clean, spacious medical examiner's office, with its laboratories full of softly humming equipment, was actually rather comforting in its matter-of-factness.

For the most part, we're shielded from the realities of what happens to our bodies after death; even the goriest *CSI: Crime Scene Investigation* or *Bones* episode is sanitized. The dead are safely unrecognizable, and holographic reconstructions and flashy animations stand in for much of the visceral reality of a forensic examination. I'm sure many people would prefer not to see the gory bits anyway, and figure they're happier not knowing.

Because you picked up this book, though, I'm guessing you're not one of those people—in which case, I have a few stories for you. The theme of death has led me to find fascinating science in many surprising places, from so-called body farms to hand-made miniature crime scenes to British Columbia, where mysterious disembodied feet have been washing ashore. We'll explore those stories and many more in this chapter—and learn more about what happens to our bodies after death. (As it happens, there's a lot of life in a body after the heart stops beating—microbial life, that is.)

Although these topics are morbid, by definition, I don't think of them as sad or depressing. Although death can be emotionally devastating for the living, it doesn't mean that morbid curiosity is a desire to be sad. Instead, it's a desire to *know*—and that desire holds the potential to lead us down a productive, if not always cheery, path. (After all, this is what drives forensic scientists to analyze death and solve crimes.) Best of all, satisfying our morbid curiosity has the ability to set our minds just a little more at ease by reassuring us that even death follows the rules of the natural world. And that's something we can all understand—if we're willing to take a look.

THE WORLD'S SMALLEST CRIME SCENES

Inside the miniatures used to train detectives

he entire Judson family is dead. Bob Judson, a foreman in a shoe factory, lies face down on a quilt beside his bed, still in his bloody pajamas. Nearby, his wife, Kate, looks like she's sleeping peacefully, except for the blood sprayed across her pillow and on the wall behind her head.

In the next room, it's even worse. Baby Linda Mae's tiny arms are thrown up beside her bloody face. In contrast with the carnage around her, Linda Mae is tucked neatly under a pink blanket decorated with dancing elephants and dogs wearing tutus.

It's a sad and desperate scene, and I've been assigned to piece together what happened. This is my first homicide investigation, and I'm nervous about my skills—despite the fact that all of the Judsons are under six inches tall and made of porcelain.

The dead family, you see, is part of a dollhouse-size crime scene built by Frances Glessner Lee, heir to International Harvester's tractor and farm equipment fortune. Lee built 20 of these eerily accurate dioramas, called the Nutshell Studies of Unexplained Death, in the 1940s and '50s. She was famous for her attention to detail, gleaning many of her models' clues from real crime scenes. They're so accurate that 18 of them are still used as teaching tools for detectives (one more was found in the attic of her New Hampshire estate in the 1990s, and another was accidentally destroyed during a move).

I'm in a dimly lit room at the Baltimore medical examiner's office surrounded by those 18 miniature scenes. The Judsons are part of the "Three-Room Dwelling," the largest of the Nutshells and the one occupied by the most dead dolls (which might be what has drawn me to it—after all, I like a challenge).

Some of the other scenes could be episodes in a home renovation show gone tragically wrong. The Pink Bathroom has a dead woman's face visible in a dressing mirror; in the Kitchen, a woman appears to have committed suicide by gas oven—or was she murdered?; and in the Dark Bathroom is another dead woman in a bathtub, plastic water frozen in time as it streams across her face.

Each scene is filled with tiny clues for observers to ferret out, and I'm joining the ranks of the detectives who have tried. This is the 73rd annual Frances Glessner Lee homicide investigation seminar, sponsored by the Harvard Associates in Police Science and held at the Office of the Chief Medical Examiner in Baltimore. Each year, dozens of detectives, correctional officers, prosecutors, and a smattering of other law enforcement types take this introductory course, occasionally tolerating journalists like me.

The seminar is a weeklong training ground for the criminal justice crowd, with 16 medical examiners and other forensic experts giving PowerPoint talks on everything from interpreting bloodstains to collecting trace evidence and recognizing different kinds of injuries. To give you an idea of the level of detail, blunt force injuries and sharp force injuries each get their own hour-and-a-half-long talk, complete with hundreds of photos of actual injuries caused by all manner of blunt and pointy things. (The photos are more gruesome for sharp force, thanks in no small part to the impalings, but blunt force is no slouch.)

In between the presentations—or as I start to think of them over the course of four days, the parade of horrible ways to die—students are assigned to teams of three or four, each with a different Nutshell to study. On the last day of the seminar, a Friday, the teams present their findings, revealing whether they believe a homicide has been committed and what clues can be gleaned from the scene. At the end, they learn whether their analysis matches up to the scenario as Frances Glessner Lee intended it. Her solutions are closely guarded secrets, as I'm reminded multiple times.

Now known as the "mother of forensic science," Lee built the dioramas as tools for teaching forensic principles to gumshoe detectives, who at the time got no training in the emerging field. She also endowed a Department of Legal Medicine at Harvard University that trained physicians and law officers from 1931 to 1966. The program was home to forensic research on everything from methods of poisoning to the analysis of gunshot residue. Today, the Nutshells remain a physical reminder of Lee's important contribution to American forensic science.

The Nutshell models were built on a scale of one inch to one foot, shrinking down details pulled from autopsy reports, police records, and witnesses but tempered with a dose of obfuscation. Sometimes, Lee changed names and dates in her scene descriptions, and she took liberties with details that weren't essential as evidence, such as wallpaper and decor.

According to Bruce Goldfarb, executive assistant to the chief medical examiner of Maryland and de facto curator of the dioramas, Lee spent as much money building some of the miniatures as a full-size house cost at the time. Goldfarb is amiable and energetic, with a gray goatee and hip black glasses. A former journalist, he landed what he calls a dream job with the medical examiner's office. When I meet him, he's finishing a book about Lee's contributions to forensic science.

I scurry to keep up with Goldfarb as he leads me around the building, popping in and out of laboratories and showing off a full-size apartment called the Scarpetta House, donated by mystery novelist Patricia Cornwell, in which gruesome crime scenes are re-created to train investigators. He doesn't slow down until we enter the dimly lit room that holds the Nutshells.

Goldfarb's tone is reverent when it comes to Lee's creations. "If there was an easy way to do something and a difficult way, she did it the difficult way," he says. (To wit, a miniature garbage can in the Three-Story Porch diorama, he posits, contains authentic trash.) He has also told me that all the figures are wearing underwear, though I can't pull any pants down to check because the Nutshells are locked behind Plexiglas.

One of Lee's rooms was constructed so that it couldn't even be viewed without taking the entire scene apart. In another, she insisted that a tiny rocking chair within should rock the same number of times after being pushed as its full-size counterpart. The plaster and lath is real, the walls have studs, and the doors are framed.

I was giddy. Finally, I was going to not only see these marvels up close, but also see them the way real detectives did, testing my knowledge of forensic science and my powers of observation to figure out what happened in each tiny, grisly scene.

My partners in miniature crime solving are Anthony Benicewicz, a Baltimore psychologist studying to become a forensic psychologist; Ayomide Oludoyi, a master's degree student in forensic medicine; Captain Mike Wahl, director of the Montgomery County Police Major Crimes Division; and Haley Nelson, a special agent at the South Carolina Law Enforcement Division. I feel supremely unqualified in their company.

Nelson, with long blonde hair and a handgun on her hip, looks like she could star in a crime TV drama. I find myself conscious of not standing too close to her weapon as we huddle around the Three-Room Dwelling and its dead family to start our investigation.

Between presentations, we rush back to our Nutshell to walk through possible scenarios, debating whether the scene might be a murder-suicide committed by the father, or perhaps a murder by an intruder. Two bedroom windows in the house are partly open, and next to the open window in baby Linda Mae's room, a small table has been knocked over. Nelson keeps coming back to the table: Could an intruder have knocked it over when he entered the house?

I'm not sure the windows are open far enough for an intruder to climb through, though. We start examining the scene room by room, reconstructing possible scenarios. A gun lies next to the kitchen table, near a pool of blood. We all take small LED flashlights from a bin on the wall and start looking for bullet holes and shell casings. The gun is long, and I'm calling it a rifle until our flashlights pick up a spray of tiny holes in a wall. "Could this be a shotgun?" I ask Mike Wahl, figuring this is a question for real police. He stares at it as closely as he can through the scene's protective box.

"It's either a rifle or a small-gauge shotgun," he says, right before we have to hurry back to the gunshot wounds presentation. I spend much of the talk reading about shotgun history on Wikipedia. Looking around the house later, we scour the rooms looking for shells.

Benicewicz and I decide to meet early the next morning in the Nutshells room to hunt for more clues. I arrive at eight o'clock to an empty room, and get a chance to be alone with the dioramas for a while. I try to imagine being a wealthy heiress in the 1940s and deciding to spend years of my life and a small fortune building tiny crime scenes. I imagine describing to a craftsman that I want him to build a miniature log cabin perfectly to scale, and then telling him to burn it down. Frances Glessner Lee did just that, and I peer in at the tiny charred body lying in a bed inside the burnt building.

When Benicewicz arrives, we find ourselves stuck. Everything in the Three-Room Dwelling seems like it could be a clue. The kitchen table is set with plates, teacups, and what looks to me like a miniature bottle of hot sauce. What meal is that for? Benicewicz starts trying to read a tiny newspaper lying on the kitchen floor, knowing Lee's eye for detail. We snap endless photos with our cell phones, then enlarge them. Eventually, he manages to read off a headline. It's about schoolchildren appearing in a local concert. It appears Lee has left us a tiny red herring.

Working without DNA or even fingerprints, the Nutshells force you to rely on the most basic elements of forensic science. These largely involve the biology of the dead and dying, along with a little Newtonian physics (the basic rules of nature that govern how blood spatters in the direction of a bullet's path, and how it pools at the lowest points in a dead body). To solve the puzzle, you must look, for example, at the tiny hair clinging to an otherwise harmless-looking blunt object, or the limb that was already stiff before a body was laid down.

My colleagues in crime solving find many such clues in the Nutshells. When we present our findings, most of the teams get close to Lee's solution. They find homicides staged to look like accidents, or like suicide, and even an accidental suicide that at first looked like a homicide.

My own team talked through scenario after scenario, pointing our tiny flashlights at each point our hypothetical killer might have taken an action. We considered where a shooter might have stood when firing the gun, and where he or she might have gone next, looking for evidence that lined up. Eventually, one scenario seemed to fit all the physical evidence.

I can't tell you what my team decided about the Judsons; I'm sworn to secrecy. But I can tell you that I volunteered to present our findings in front of the class, going into too much detail as is my weakness. And I can tell you that we nailed it.

If you'd like to try for yourself, you can. The Nutshells aren't open to the general public, but as part of an exhibit in 2017, the Smithsonian Institution commissioned detailed photographs of the Nutshell interiors that are available online.[*] You

---

[*] Go to *americanart.si.edu/exhibitions/nutshells*.

can even explore a virtual-reality version on your phone. You won't find the solutions to the scenes, but you shouldn't let that stop you.

Examining the Nutshells isn't just about finding the right answer, after all. The point is to learn how to really see what's in front of you—to walk into a scene where something horrific has happened and to resist the urge to look away. To experience the Nutshells is to immerse oneself in uncertainty; just as in real life, you're never quite sure what's important and what's not. That uncertainty is what makes these scenes credible. And it's what will push you to test your own deductive powers against your greatest fears.

THE LIVING DEAD

When microbes turn the tables on us

A t any given moment, roughly the same number of microbes are living within you as there are human cells in your body.* A whole lot of these tiny hangers-on are just waiting for you to drop dead—and what happens next is providing groundbreaking clues for figuring out when we die.

Within about four minutes of your demise, bacteria inside you will start cranking up the party like it's 1999. First, thanks to a process called autolysis, or self-digestion, your cells start popping open like champagne bottles. Bacteria previously relegated to fighting over scraps in your intestines suddenly find themselves awash in the microbial equivalent of a Las Vegas buffet's bottomless brunch. As they start to slurp up the cellular juices inside you, some of the bacteria on your skin start working on you from the outside in. Thus begins the process of decomposition.

It's a little unsettling, when you think about it. OK, maybe a lot unsettling. But it still begs the question: What keeps all those bacteria from decomposing you while you're alive?

....

* You may have heard that the microbes inside us outnumber our own cells by 10 to 1. It's the most famous fact about the human microbiome—but it's also wrong. The 10-to-1 ratio traces to a 1970 paper by microbiologist Thomas D. Luckey, who estimated that each gram of human feces and intestinal fluid contains about 100 billion microbes, and that an adult contains a kilogram of this material, yielding a total of 100 trillion microbial cells. Years later, microbiologist Dwayne Savage combined this with a wild estimate of 10 trillion cells in the human body, and the ratio was born. In 2016, biologist Ron Milo and colleagues debunked that ratio, projecting more like 1.3 microbial cells for each human cell. And that can vary. "Each defecation event," they write, "reduces the content [of microbes] by a quarter to a third."

That's silly, you say. I'm alive. Only dead things decompose.

Fair enough—but why is that? One crucial reason is that when you're alive, your immune system fends off decomposition by waging a battle to the death with microbes. Our bodies are under constant assault from bacteria, fungi, and viruses that would love to get inside us, consume our delicious organic matter, and reproduce like mad. We call these microbes germs, and when they harm us—whether by eating us alive, hijacking our cells to replicate themselves, or making toxic waste products—we call that an infection. Our immune system is on call 24/7 to fight off these invaders, while giving a pass to more benign bacteria, like the ones in our guts that help us digest food. It's a pretty amazing feat.

Decomposition, on the other hand, is what happens when you die and your immune system shuts down. At that point, your body essentially gets its first break from a war it has been fighting every moment of your life. The bloodstream no longer delivers oxygen and nutrients to tissues and organs, and cells begin to die. Meanwhile, enzymes start to dissolve dying cells from the inside out, eventually popping those cellular champagne bottles. That happens quickly in the pancreas and liver, which are rich in enzymes, and in the brain, which contains abundant water. Bacteria begin to run amok, taking over organ after organ as they multiply.

Though most of us may prefer not to think about our eventual decomposition in this kind of detail, forensic scientists have found it useful to document exactly which bacteria make a meal of our dead bodies, and to time their activities. They've even given this motley community of scavengers a name: the necrobiome. By mapping out the sequence of microbes that come and go during decomposition, they hope to estimate more accurately how long a body has been dead.

Despite technological advances, this has remained one of the trickiest questions in forensic science. Medical examiners are often loath to give anything but a broad range for time of death, using classic signs like lividity (blood pooling after death), insect activity, or stage of decomposition. The longer a person has been dead, the larger the ranges are.

Until recently, we knew surprisingly little about which microbes ultimately consume us, or the tiny dramas that play out as they battle for corpse dominance. In 2013, a Colorado research team broke new ground by reporting that the microbial community living on dead mice changed in a consistent pattern over time; this meant that scientists might be able to narrow down a time of death by observing how the microbes on a dead body compare to that pattern.

It's the same principle as forensic entomology, which studies the insects on a dead body. Flies are typically quick to land on a cadaver and lay eggs, for example—and those eggs grow into maggots and then metamorphose into adult flies in a predictable time line. So if you measure the maggots of a particular fly species, you can estimate how long the person has been dead. With bacteria, the same principle is applied by swabbing key body parts, using DNA to identify the types and abundance of various bacteria, and observing how that composition compares to a typical time line of decay.

The concept shows promise, thanks to advances in managing enormous data sets of bacterial DNA. Studies of cadavers have displayed a fairly predictable pattern of bacterial growth as humans decompose, because different bacteria prefer different entrées on their own Las Vegas buffet—that is, a dead human body with its variety of tasty offerings.

At the only Vegas buffet I've been to, I found myself wandering in awe amid piles of fried chicken and shrimp, mentally calculating the best value in price and deliciousness given my limited stomach size. Should I dive into the crab legs first, or go straight for the prime rib? Similarly, the microbes that consume us have their own preferences and order of operations. Some flourish in the mouth, whereas others focus on the other end of the gastrointestinal system. And some are able to work in an anaerobic environment, without oxygen, allowing them to get a jump on the nutritious internal organs before they've been exposed to air.

But to turn all this into a useful time line of human decomposition, someone has to swab a lot of dead bodies. The first to give it a go were researchers working at the Southeast Texas Applied Forensic Science facility—one of a handful of U.S. anthropological research sites, often called body farms, where scientists study the decomposition of human cadavers under a variety of conditions. In 2013, they cataloged bacteria growing inside two cadavers decomposing under natural outdoor conditions, focusing on the bloat stage.

Bloat is what it sounds like. During decay, bacteria produce gases such as hydrogen sulfide (which stinks) and methane (which does not, despite many body function jokes to the contrary). These gases inflate the cadaver and eventually force fluids out in what's called a purge, or rupture event. That's just as gross as it sounds, but it's an important stage of decomposition that forensic scientists and medical examiners have traditionally used to get a rough estimate of how long someone has been dead. Depending on temperature and other conditions, for instance, a bloated body may have been dead somewhere between two and six days, although the purge might happen anywhere from five to 11 days after death.

Researchers sampled the mouths and rectums of bodies at the Texas body farm—areas you'd expect to be rich in bacteria— before and at the end of the bloat stage. Next, the team identified the bacteria in these samples by looking for characteristic genetic markers. By the end of the bloat period, they found anaerobic bacteria such as Clostridia had become dominant.* And mouth scrapings of both bodies showed a shift during bloat toward Firmicutes, a group of bacteria that includes Clostridia.

----

* Scientists who sampled the livers and spleens of 45 cadavers in Alabama have even given a name to the proliferation of these bacteria after death: the postmortem clostridium effect.

This was the first sign of a potentially useful pattern of bacterial growth on cadavers.

Those first two bodies were just the start, though. To find a pattern predictable enough to use as a time line after death, scientists would need to sample a lot more cadavers under a variety of conditions. And that's what Jennifer Pechal, a forensic scientist at Michigan State University, set out to do.

As of 2019, Pechal had samples of microbes from nearly 2,000 human cadavers. I watched her give a presentation at the American Academy of Forensic Sciences meeting and witnessed just how far the research in the strange little field of death bacteria has progressed. Pechal now works closely with the Wayne County Medical Examiner's office, which routinely swabs bodies arriving at the morgue. A handful of other research groups across the United States and around the world have also jumped on the bandwagon: Scientists in France, Austria, and Italy are comparing their data, and finding that no matter where we live— or die—our necrobiomes follow similar patterns.

The picture that's emerging so far is that there's a large, consistent shift in the microbial community after about 48 hours. With that marker, it's fairly easy to tell whether a body had been dead for more or less than two days, based on its microbial decomposition. Pechal hopes continuing microbial research will be able to narrow down the time since death to a 12-hour window or less. "I'm optimistic that this will be something that medical examiners can use in the future," she says.

More surprising, the team also discovered that bacteria found in the mouths of the recently deceased may hold clues to their health before they died. The same bacteria associated with heart disease in the living were found in the mouths of people with heart disease who had been dead for up to 24 hours. Based on this research, Pechal says it's possible that one day, a microbial test of a dead body could help give pathologists an idea of not only how long the person has been dead, but even the cause of death—for example, whether undiagnosed heart disease could be a factor.

Next, Pechal says the scientists are focused on developing computer models that can analyze a dead body's microbial passengers and estimate the postmortem interval, or time since death. But before medical examiners can use those models, they will have to be tested using bodies with a known time of death to ensure they're accurate. The hope, she explains, is ultimately to provide a "living tool" for medical examiners and investigators—one that, in combination with other methods, can become a reliable way to narrow down a postmortem interval.

That said, this scenario is still five or 10 years away. First, there are a lot of practical issues to address—for example, finding the most reliable spots on a cadaver to collect bacteria. "Do you use the outside? Do you use the inside? Do you use a combination of both?" Pechal asked at the outset of the work. Based on data from the first 188 cadavers the team has analyzed, mouths and ears are good sampling locations for bodies that have been dead on the scale of several days. Those are also fairly easy spots to sample, and Pechal hopes that one day, swabbing for microbes might become a routine part of an autopsy.

The bacteria that consume dead bodies can also leave behind other clues to time of death. Not only can we look for which bacteria are on a body and how abundant they are; we can also detect signature compounds that they make as they decompose our flesh.

Back in the 1990s, forensic anthropologist Arpad Vass decided to catalog the chemical compounds that this large cast of characters produce as they break down a body. He found three compounds that are useful for estimating time of death: all fatty acids produced by the decomposition of human fat, muscle, and the food left in our guts after we expire. Since then, these and other biochemical compounds have been used in forensic investigations; in one case that Vass has written about, fatty acid tests of a man found dead in the woods suggested he had died 52 to 57 days earlier. Police quickly searched missing persons records from around that period and were able to identify the man, who was later found to have died 50 days before his discovery. The fatty acids, as it turned out, came pretty close.

Even a dead person's DNA could hold clues to the postmortem interval, because it appears that some gene activity persists after death. Tom Gilbert, a geneticist at the Natural History Museum of Denmark, told me it's like boiling pasta; if you turn off the heat, the water keeps bubbling, just slower and slower over time.

Likewise, genes in some parts of the body keep genetically percolating longer than others. Combining the most predictable tissues, Gilbert was able to estimate a time of death in his lab experiments that was accurate to within an astonishing nine minutes during the first few hours of death. After longer time periods, he notes, the genetic information starts to become too degraded, and bacterial changes become the more useful tool.

These various chemical changes are promising for improving time of death estimates in the near term, says Jeff Tomberlin, a forensic entomologist at Texas A&M University. Tomberlin has studied the progression of both kinds of "bugs," insects and microbes, on cadavers. As he puts it, there are so many kinds of microbes, and they vary so much genetically, that cataloging them all and seeing how they grow in various environments is going to take time. But eventually, all these approaches could work together. "The unseen world is going to be the future of forensic science," Tomberlin observes. And perhaps one day, it will be as routine to scan a body's microbial and chemical fingerprints as it is to ink its actual fingerprints.

WOULD YOUR DOG EAT YOU IF YOU DIED?

Quite possibly (but you'll be none the wiser)

In 1997, a forensic examiner in Berlin reported in the journal *Forensic Science International* on a particularly grisly case. A 31-year-old man had retired for the evening to the converted garden shed behind his mother's house, where he lived with his German shepherd. Around 8:15 p.m., neighbors heard a gunshot.

Forty-five minutes later, the man's mother and neighbors found him dead of a bullet wound to the mouth, a Walther pistol under his hands and a farewell note on a table. And then police made an even more gruesome discovery: bite marks on his face and neck.

The man's German shepherd was calm and responded to police commands. But on the way to an animal sanctuary, the dog vomited some of its owner's tissue, including skin with still recognizable hair from the man's beard. This wasn't a situation where the animal was trapped with nothing else to eat; a half-full bowl of dog food was sitting on the floor when police arrived. The disturbing implication: Maybe man's best friend isn't so loyal after all.

No one tracks how often pets scavenge their expired owners' bodies. But over the past 20 years or so, dozens of case reports have appeared in forensic science journals, providing the best window we have into a possibility that most pet owners don't like to contemplate: becoming pet food. It's not something I had ever considered, until a friend asked me about the phenomenon. As an animal lover living alone, she was maybe a little concerned about her prospects.

So I set out to learn as much as I could. Eventually, I found studies describing more than 20 cases of people scavenged by pets, as well as a 2015 study that documented a total of 63. The stories are as tragic as the photographs are grisly, and I've read them in an attempt to answer the question: Why? Why do some pets starve for weeks without touching their dead owners, while others tuck in right away? And is there any way for pet owners to avoid such a grisly end?

Some of the patterns I found as I pored through these papers are surprising, and they raise questions about why pets might be motivated to eat the dead. They also illustrate just how wrong we can be in interpreting animals' behavior when we fail to see things from their perspective.

First things first: Cats get a bad rap. People tend to think of dogs as loyal, while their feline counterparts are considered aloof predators who'd eat you as soon as they'd look at you. But as it turns out, most animal scavenging documented in the forensic science literature involves dogs. In fact, I was able to find relatively few published accounts of cats eating their owners (and one of these was a man who had 10 cats). In one report, published in the *Journal of Forensic and Legal Medicine* in 2010, a woman had died of an aneurysm and was found the next morning on her bathroom floor. Forensic testing revealed that her dog had consumed much of her face, while her two cats hadn't touched her.

When I first wrote about this phenomenon on the Gory Details blog, I theorized that medical examiners might be more likely to report cases involving dogs because they find the behavior more noteworthy in man's supposed best friend. But after talking to forensic scientists and medical examiners, I'm not so sure.

As a 2016 study documented in *Journal of Veterinary Behavior* observed, "Canine scavenging in indoor settings is rarely reported, but is regularly observed in forensic practice." The medical examiners I've talked to confirm this. Joseph Prahlow,[*] a medical examiner in Michigan, said he had seen evidence of pet predation on a corpse during an autopsy "at least a couple times a year." Usually, he said, dogs, rather than cats, have done the scavenging.

That makes sense when we look at feeding behaviors of dogs versus cats. Generally, dogs are opportunistic eaters; they hunt, but are also scavengers not above nosing around in a dead squirrel. Cats prefer to hunt and kill their prey. Although neither is above digging through garbage if hungry, dogs tend to be less picky about eating whatever they come across.

"Dogs are descended from wolves," says Stanley Coren, a psychologist who has written books and hosted television shows about dogs. "If we have a situation where the owner dies and there's no source of food, what are they going to do? They're going to take whatever flesh is around."

This isn't to say that you can feel entirely confident your cat won't eat you. When cats do scavenge human remains, they tend to go for the face—especially soft parts like the nose and lips, says forensic anthropologist Carolyn Rando of University College London, who took an interest in the topic after studying scavenging at archaeological sites. "It doesn't surprise me, as a cat owner," she says. "If you're sleeping, they tend to swat your face to wake you up." So a cat might start out trying to "wake up" a dead owner, and then begin to bite when that doesn't work.

In some forensic reports, it's clear that animals were scavenging their dead owners to survive. In a 2007 account, a chow chow and a Labrador mix lived for about a month after consuming their dead owner's body, leaving only the top of the skull and an assortment of bone shards.

And yet in the 1997 Berlin case, the German shepherd began eating parts of its owner soon after death. "It is interesting to

---

[*] As it happens, I had just seen Prahlow give a particularly horrifying presentation at a forensic science conference about elevator-related deaths. The takeaway: Don't ever try to get out of a stalled elevator.

consider the reasons for an otherwise well-behaved pet with no motivation of hunger to mutilate the dead body of its owner so quickly," wrote Markus Rothschild, the forensic examiner. Many people assume that a dog would only eat its dead owner if it were starving. "However," Rothschild writes, "forensic experience shows that this is clearly wrong."

In the 2015 review of scavenging by dogs, nearly a quarter of the cases featured pet owners who were partially eaten within less than a day of dying. What's more, some of the dogs—like the one in Berlin—had access to other food they hadn't eaten.

And if you think about it, if dogs only ate their owners because they were starving, you might expect them to consume the same body parts that they prefer in the wild. But they don't. When pet dogs scavenged dead owners indoors, 73 percent of cases involved bites to the face; just 15 percent showed bites to the abdomen. By contrast, canines scavenging outdoors (both coyotes and domestic dogs) have a well-documented pattern, opening the chest and abdomen to eat the nutrient-rich organs early on, followed by the limbs. Only 10 percent of outdoor scavenging involves wounds to the face or head.

It's tempting to think that if you're close to your dog and have treated her well, you're probably off the hook if you die in her presence. But canine behavior isn't quite so clean-cut. None of the reports I saw indicated any prior history of animal abuse—and in fact, several noted that the owners had very good relationships with their dogs, according to friends and neighbors.

Instead, consider a pet's psychological state: "One possible explanation for such behavior is that a pet will try to help an unconscious owner first by licking or nudging," Rothschild observes. "But when this fails to produce any results, the behavior of the animal can become more frantic and in a state of panic, can lead to biting." From biting, it's an easy jump to eating, Rando observes: "So it's not necessarily that the dog wants to eat, but that eating gets stimulated when they taste blood."

Further complicating matters, Rando adds, is that different dog breeds have different temperaments, which could play a role in how they respond to an owner's death. But as it happens, many kinds of dogs turn up in forensic reports of scavenging, including lovable Labs and golden retrievers. The accounts I read involved a mix of mutts, as well as several hunting or working dogs.

Overall, most of the dogs were medium to large, with a beagle being the smallest breed I've seen reported scavenging. However, because larger, more powerful breeds can inflict more damage, those cases might be more likely to be highlighted. In 2011, European scientists reported three scavenging cases where cadavers had been eaten to the point of decapitation; all involved German shepherds. Still, for all we know, a Pomeranian or Chihuahua would bite your head off if it could.

Rando suspects that an individual dog's temperament might matter even more than its breed. For example, an insecure, fearful dog that regularly shows signs of separation anxiety may be more likely to transition from frantic licking to biting to eating.

So, what can you do to avoid being eaten? There's no way to guarantee it won't happen, regardless of what type of pet you have; even hamsters[*] and birds have been known to scavenge on occasion. The best way for pet owners to reduce the odds, Rando says, is to make sure people will stop by if they don't hear from you. By the same token, she advises, make sure to check in on neighbors who are elderly, sick, or vulnerable. "It's a good reason to make sure you have people around you," she says. "Social activity later in life is good for everybody."

Then again, not everyone is too worried about this gruesome scenario. In fact, when this blog post was first published, I was surprised by how many people were actually fine with it.

....

[*] In fact, the hamster case was quite nasty. A 43-year-old woman in Germany was found dead in her apartment with wounds to her face and head, and at first investigators thought she had been attacked. But then they looked in a drawer and found her pet golden hamster's nest, lined with shreds of its owner's skin, fat, and muscle tissue.

"You're providing for your pets after death. That's good!" one reader tweeted.

But even if you're not so sanguine, you might consider giving dogs (and cats) a pass here. The fact that they sometimes nibble on our dead bodies even before they're hungry doesn't necessarily mean they're indifferent or see us as a walking can of Alpo. In fact, dogs' distressed attempts to rouse their deceased owners suggest that losing a human companion is a traumatic experience for them.

And in the face of trauma, we obviously can't expect our pets to behave like humans in mourning; it makes more sense to examine how other social animals react to death. Although it's impossible to know whether a dog's emotional state would be recognizable to us as grief, studies of animal behavior suggest that many species respond in complex ways to the death of their own kind. Elephants will run the tips of their trunks along the face and tusks of a dead compatriot—the same body parts they would have touched in greeting during life. Chimpanzees, dolphins, and dingoes (wild dogs) may carry dead infants around for days or weeks. And crows, as we'll discuss later, gather noisily around their dead—and sometimes even attack the corpse.

In the end, dogs' extraordinarily close bond with humans, developed over generations of domestication, may actually make them more likely to eat us. Think about how distressing it would be to find your human buddy laid out on the floor, unresponsive. So if that distress leads to licking, and then eating, I guess we'll just have to accept the fact that sometimes, love bites.

THE CORPSE THAT BLED

Can the dead reveal their killers?

I f only the dead could talk. If we could just extract some sign from a cadaver—maybe an image of a pouncing killer burned into the retinas of a murder victim—it would save forensic medical examiners and police a lot of effort.

To that end, the criminal justice system has tried some quacky science over the years to obtain such clues. As recently as a century ago, for instance, some doctors were plucking out the eyes of corpses and trying to develop the retinas' last images, much like photographs. The process was called optography, and although it turned out to be bunk, it was premised on a real scientific discovery. In 1876, German physiologist Franz Christian Boll found a purplish pigment in the back of the eye that pales in response to light; he called it "visual purple," and today it's known as rhodopsin. Alas, the "optograms" collected by fixing rhodopsin in corpses' eyes were meaningless, and never revealed any killers.

Although optograms were a relatively short-lived fad, another way of interviewing the dead persisted much longer. Until just a few centuries ago, people could be convicted of murder based on the idea that a corpse would spontaneously bleed in its killer's presence. From at least the 1100s to the early 1800s, men and women were judged in courts across Europe and colonial America based on a test called cruentation, or the ordeal of the bier, named for the type of wagon that carried a corpse or coffin. In such testimony, oozing knife wounds and gushes of blood from the noses and eyes of the deceased were considered proof positive of guilt.

No one knows exactly how the belief in cruentation got its start. But one of the earliest mentions on record comes in the 13th century, in the epic Germanic poem *Nibelungenlied*. In it, the dragon slayer Siegfried is murdered, and his body is laid out on a bier. When his killer, Hagen, approaches, the dragon slayer's wounds begin to flow.

"It is a great marvel, and frequently happens today that whenever a blood-guilty murderer is seen beside the corpse, the wounds begin to bleed," according to the anonymous poet who wrote *Nibelungenlied*. The words make it clear that by the time the poem was written sometime around the year 1200, the idea of cruentation had already caught on.

Applying even the most rudimentary science, it's hard to imagine how the idea of corpses bleeding on cue arose. Even setting aside the fact that dead bodies have no way to sense a murderer's presence, simple observations of cadavers would seem to rule out spontaneous bleeding. For instance, once the heart has stopped, a dead body doesn't really bleed. Soon after death, blood begins to settle to the lowest parts of the body, a process called livor mortis. The settled blood imparts a purplish blue hue, or lividity, that generally becomes visible from 30 minutes to four hours after death. The lividity becomes "fixed" within about eight to 12 hours, meaning that even if a body is moved, the settled blood will stay in place. "During this time, the body won't really bleed; it might ooze," says A. J. Scudiere, a forensic scientist who also writes crime novels.

So, what did people see that convinced them that a cadaver had suddenly started to bleed? It's possible that if a body had started to decompose, they may have observed a foul-smelling red-brown liquid called purge fluid that can build up in the lungs. If a body brought forth for trial was poked or jostled, some of this fluid could have leaked from the nose or mouth.

Purge fluid may have lent occasional credibility to cruentation. But then again, people weren't necessarily looking for physical proof; they believed in literal courtroom miracles. The ordeal of the bier was just one of several divine interventions used as tangible evidence in courts. There were also ordeals by water, including the famous test in which witches float and the innocent sink. In ordeals by fire, suspects were forced to hold or walk over hot iron; they were deemed guilty if God didn't heal the wounds within three days. Such trials weren't confined to small towns or backwater provinces: Even King James I of England was a firm believer in cruentation.

King James is more famous today for his version of the Bible than for his belief in diabolical practices. But in 1597, more than a decade before the King James Bible was published, he wrote a treatise on demons and sorcery called *Daemonologie, In Forme of a Dialogue*. The king was obsessed with the occult, and with witches in particular, having flushed out a coven of at least 70 in 1590, when he was still known as King James VI of Scotland. The witches were tortured using devices like the "breast ripper"—which is exactly as horrific as it sounds—until they confessed. Eventually, some 4,000 people were burned at the stake in Scotland's witch trials.

In *Daemonologie*, the king wrote of his belief in cruentation as a way to mete out justice:

> In a secret murther, if the deade carcase be at any time thereafter handled by the murtherer, it will gush out of bloud, as if the bloud were crying out to the heaven for revenge of the murtherer, God having appoynted that secret super-naturall signe for tryall of that secret unnatural crime.*

....

* Later in the same sentence (King James loved a run-on) he explains the ordeal by water for detecting witches: "God hath appoynted (for a super-naturall signe of the monstruous impietie of the Witches) that the water shal refuse to receive them in her bosom." He also notes of witches that "Not so much as their eyes are able to shed teares (thretten and torture them as ye please)," which he did as well. As for women who did cry, James said their tears were shed "dissemblingly like the Crocodiles."

Oddly though, it seemed that men's dead bodies generally had the most crying out to do. In her master's thesis, historian Molly Ingram of the University of Oregon examined accounts of cruentation, many from early pamphlets and broadsheets describing murder trials. Notably, women were rare in accounts of bleeding corpses—except as the accused killers. Women's testimony was also largely missing in accounts of court proceedings. "Female speech was considered less credible than male speech," Ingram says.

Ingram also studied historical records describing possession by demons. (This was thought to happen mostly to women, whose bodies were presumed to be weaker than men's and thus more easily invaded.) Ingram found that the real women's speech was sometimes trusted less than that of the male demons who supposedly possessed them. Given the misogyny of the times, "I don't think it's surprising that there was a difference," Ingram observes. "What was more surprising was that no one seemed to notice this or talk about it [today]."

In one rare account of the ordeal of the bier being applied to a woman, a Maryland man named Thomas Mertine was accused of beating his female servant Catherine Lake to death in 1660. "There was noe issue of bloud from the Corps," the court stated, confirming what the jury seemed to have already decided: Despite the testimony of three servants who saw Mertine beat his servant to death, the jury found that Lake had died not from the beating but of an ailment called "fits of the mother," akin to hysteria. The master walked free.

Even into the early modern era, when Christopher Columbus reached the New World and the Renaissance blossomed, people still relied on magic and miracles to inform legal disputes. As Ingram puts it, "The world remained an enchanted place."

Although most forms of trial ordeal died out in the 16th century, cruentation hung on for a while longer. Ingram suspects it may have been more trusted than other ordeals because of its association with men, rather than women.

Perhaps one of the last uses of cruentation occurred in Lebanon, Illinois, in 1869. The historian Henry Charles Lea wrote in 1878 that "the bodies of two murdered persons were dug up, and two hundred of the neighbors were marched past them, each of whom was made to touch them in the hope of finding the criminals." I haven't been able to get my hands on the original newspaper account in the March 29, 1869, *North American and United States Gazette* to find out what happened next, but presumably the body didn't give up the murderer.

Today, thank goodness, you'll see "talking" corpses only in art and plays. At the start of Shakespeare's *Richard III,* for instance, the hunchback Richard (then Duke of Gloucester) has killed King Henry VI. Richard's future wife, Lady Anne Neville, accuses him of this treachery when he approaches her on the way to bury the king, and the corpse begins to bleed.

> Oh Gentlemen, see, see dead Henries wounds, Open their congeal'd mouthes, and bleed afresh. Blush, blush, thou lumpe of fowle Deformitie: For 'tis thy presence that exhales this blood.

Today, rather than relying on testifying corpses and wounds, we rely on DNA, fingerprints, and other forensic methods to help determine guilt or innocence. In fact, scientific evidence is now so important in murder trials that lawyers fear juries will expect high-tech evidence in every case; they call it the "CSI effect." It's not yet clear, based on studies of jurors, whether this effect is real. But evidence shows that fans of shows like *CSI* may evaluate evidence differently from others; a 2015 study by Canadian criminologists demonstrated that mock jurors who said they believed the forensic TV programs they watch are realistic put more weight on DNA evidence than did their counterparts.

So although what we believe in has changed over time, our biases still follow us into the courtroom. Jurors may not expect the spectacle of a corpse bleeding from its wounds, but they still enjoy a little forensic razzle-dazzle, like that "gotcha" moment when an expert reveals a DNA match with billion-to-one odds. And when you think about it, that's not so different from our ancestors hoping to get word from beyond the grave; today, the message just takes a more scientific form.

Maybe it's no surprise that death stokes our imaginations and makes us think less than objectively about the evidence before us. After all, it remains the great unknown.

IF THE SHOE FLOATS . . .

A flood of feet in British Columbia

On August 20, 2007, a 12-year-old girl spotted a lone blue-and-white running shoe—a men's size 12—on a beach of British Columbia's Jedediah Island. She looked inside, and found a sock. She looked inside the sock, and found a foot.

Six days later on nearby Gabriola Island, a Vancouver couple enjoying a seaside hike came across a black-and-white Reebok. Inside it was another decomposing foot; it, too, was a men's size 12. The two feet clearly didn't belong to the same person; not only were the shoes themselves different, but they both contained right feet.

Police were stunned. "Two being found in such a short period of time is quite suspicious," Garry Cox of the Royal Canadian Mounted Police told the *Vancouver Sun*. "Finding one foot is like a million to one odds, but to find two is crazy. I've heard of dancers with two left feet, but come on."

The next year, five more feet appeared on nearby Canadian beaches. The discoveries ratcheted up the public's fears, and media speculation soared. Was a serial killer on the loose? Did he have something against feet?

Over the course of the next 12 years, a total of 15 feet washed ashore in the area around Vancouver Island, a network of waterways called the Salish Sea. Six more turned up in Puget Sound, which lies across the U.S. border at the southern end of the sea. With the exception of one foot wearing an old hiking boot, all of them were encased in sneakers. The sneaker-clad feet became famous, even garnering their own Wikipedia page. And with fame came hoaxes: pranksters stuffed shoes with chicken bones or skeletonized dog paws and scattered them along Canadian shorelines.

Tipsters called police with all manner of theories about the origins of the feet. "We get some very interesting tips that come in about serial killers, or containers full of migrants that are sitting at the bottom of the ocean. Aliens—had that one as well," says Laura Yazedjian, a forensic anthropologist who works as a human identification specialist for the British Columbia Coroners Service. "And occasionally a psychic. Actually, pretty much every single time, a psychic will call and offer to help."

But this type of mystery, it turns out, requires scientific, rather than criminal investigation (or psychics). In fact, science can answer all of the obvious questions—for example, why are feet, and not entire bodies, washing ashore? And why are they showing up on this particular stretch of British Columbia's shores? But the research that has addressed these questions is anything but obvious. To understand how the feet got where they did, we have to follow some unexpected lines of inquiry, involving everything from the science of sinking to the decomposition of pigs and spreading oil spills.

To begin, we must understand what happens to a dead body once it's in the water. So let's follow the adventures of a seafaring cadaver.

Once in the water, a cadaver's first move will be either to float or to sink. This is a surprisingly crucial step, as it will help determine what happens next. A floating object will be carried with the winds and by surface currents, and might soon wash ashore. A sinker, on the other hand, might remain in place, or be tugged in a different direction by deeper currents. What's more, a floating body, exposed to air, will decompose differently from one that sinks, with ramifications for the fate of its feet.

One might assume that a drowned person will sink because their lungs are full of water, and that a cadaver's air-filled lungs would otherwise act as a flotation device. But the reality is not so simple. Using data collected in 1942, E. R. Donoghue of the Armed Forces Institute of Pathology set out to settle the matter in a 1977 article titled "Human Body Buoyancy: A Study of 98 Men." The 98 men in question were "healthy U.S. Navy men in the 20-to-40-year age group." Each was suspended underwater and weighed both with his lungs full of air, and after expelling as much air as possible. It's no easy task to wait to be weighed underwater with no air in your lungs—but again, these were Navy men.

With their lungs fully inflated with air, all the men floated. But once they had emptied their lungs (as would be the case with a dead body) most of the men sank in freshwater; only 7 percent floated. In seawater, though, people are more buoyant: 69 percent of the Navy men would float if they were dead and naked in the ocean, Donoghue estimated. But it was a close call; just a little added weight, such as heavy clothing or water in the lungs, could cause a body to sink. In the end, the data suggest, cadavers are overall more likely to sink than to float, and people who drown are the most likely to sink.

What's more, once a body sinks, it tends to go straight to the bottom. Sometimes, an underwater cadaver will eventually bloat, just like a body on land, causing it to bob to the surface. But that doesn't always happen, says Yazedjian, the investigator from the Coroners Service. In a deep lake or ocean, it may never come back up. Not only does the cold inhibit decay in deep waters, but the greater water pressure there also prevents any gases from expanding and causing bodies to float. Instead, other microbial processes take over and convert a sunken body's tissues to adipocere, "this kind of waxy, soaplike tissue," she says. Adipocere can persist for years, even centuries, in a low-oxygen environment.

And that's exactly what Yazedjian saw on the feet she examined from the Salish Sea. They were covered in adipocere, suggesting that the cadavers sank, and remained underwater as they decomposed. That could explain where the remainders of the bodies were: They sank and stayed sunken.

But why didn't the feet stay down with the bodies?

To understand how the feet set sail sans bodies, we need to know how a human body might decompose underwater, and whether its feet are prone to pop off and float away. Scientists study the process of human cadaver decomposition at several U.S. forensic research sites, but these are all on land; none had ventured to drop a body into the ocean.

But our investigation is not dead in the water. In the summer of 2007, forensic scientist Gail Anderson of Simon Fraser University was conducting a study for the Canadian Police Research Centre to understand how quickly a homicide victim would decompose in the ocean. Because ethics rules preclude using a human body, she used a dead pig instead. Pigs have often been used in forensic research as stand-ins for a human body; they are roughly comparable in size and are quite similar biologically.

Even better, Anderson conducted her study in the Salish Sea, not far from where the third human foot would be found six months later. Her team dropped the dead pig into the water, and it promptly sank 308 feet to the seafloor. What happened next was not pretty. The pig carcass was quickly eaten by a large and unruly mob of shrimp, lobsters, and Dungeness crabs, starting with the "expected areas, the anus region and the facial orifices," Anderson reported. It was as if a Red Lobster buffet had risen up to exact its revenge.

Since then, Anderson has dropped more pigs even deeper in the Strait of Georgia, a main channel of the Salish Sea, and found that in some cases scavengers can skeletonize a carcass in less than four days.

So what about the feet? It turns out that underwater scavengers like crustaceans will work around bones and other tough obstacles, preferring to pick apart softer tissues. And unlike the bony ball-and-socket joints that join our legs to our hips, our ankles are made up mostly of soft stuff: ligaments and other connective tissue. So it follows that a sunken, shoe-wearing cadaver in the Salish Sea is likely to be chewed apart by scavengers, and to have its feet disarticulated from the rest of the body in short order.

And as Yazedjian tells me, all of the Salish Sea feet appeared to have been separated from their bodies by natural processes, like scavenging and decomposition. "Please don't call them 'severed feet,'" she warns. Severed means that someone cut them off, she explains, and the Coroners Service never found cut marks on any of the bones to suggest that.

What's more, feet wearing sneakers made in the last decade or so would almost certainly float. Not only have gas-filled pockets become common in sneaker soles (and they're visible in some sneakers found in the Salish Sea), but around that time, the foams used in sneaker soles started to be noticeably lighter, with more air mixed in. In other words, they've become buoyant.

So now we have a seafaring foot, sneaker-clad and ready to sail. But why the Salish Sea? If feet are likely to float away from dead bodies, why aren't beaches everywhere littered with them?

Possibly the man who knows the most about how and where things end up in the Salish Sea is Parker MacCready, a professor of oceanography at the University of Washington in Seattle. He's built a three-dimensional computer simulation of the coastal ocean of the Pacific Northwest, including the Salish Sea. "It's all realistic," he says, "in the sense that it has realistic tides, winds, rivers, and ocean conditions." The simulation is called Live Ocean, and as we talk on the phone, we both watch it running on his website: Brightly colored water sloshes around a map according to that day's weather and tides.

MacCready uses the model to predict where an oil spill would travel over the course of three days. As we watch, black blobs appear near Seattle and Tacoma, simulating the hypothetical oil spill, and immediately start flowing north into Puget Sound, sailing on rainbow-colored swirls that depict moving waters of various salinities. Soon, the blobs break apart into thin streamers and dots, splitting and sloshing in every direction as tides and currents push them around.

As it turns out, Live Ocean reveals an important key to the mystery—why so many feet are washing up here in particular. The answer? The Salish Sea has the perfect storm of feet-ensnaring properties.

The reasons add up. First, it's an unusually large and complex body of inland water, which acts as a trap. As MacCready's model shows, once something goes in the water, it might wash ashore in plenty of places—but it's still within the Salish Sea. Second, the prevailing winds are easterlies, so they bring stuff in from the ocean, rather than pushing it out to sea. And finally, there's something MacCready's model doesn't show, but he points it out. You see a lot of folks wearing sneakers at the beach in the Pacific Northwest, where many choose to hike among the slippery rocks. Taken together, all these factors—plus the cold deep waters and healthy scavenger populations—make the Salish the ideal foot magnet.

But who were the owners of the Salish Sea feet? The first place investigators looked was missing person reports. The Coroners Service has now compared DNA from each foot to a database of more than 500 missing people in British Columbia, plus Canada's new National Missing Persons DNA Program, launched in 2018.

Using DNA, the team linked nine of the feet to seven missing people. (For two, both feet were found; most had been missing for a year or more.) The longest-missing person had disappeared in 1985; his foot in a hiking boot was found in 2011. In the most recent case, the foot of a young man who disappeared in 2016 was documented to have washed up on an island in Puget Sound in 2019.

The Coroners Service in British Columbia reports that none of the Canadian cases so far have been found to result from homicide. In some cases, it became clear that the person had died by accident or suicide, as in the case of one woman who jumped from a bridge. Other times, circumstances were hazier. In the case of a young man whose foot was found in Puget Sound in 2019, U.S. police said they couldn't rule out either homicide or suicide. And for those who vanished without witnesses, it's nearly impossible to glean a cause of death from a foot alone.

As of this writing, five of the feet in British Columbia remain unidentified.

Some, no doubt, will be disappointed to learn that a serial killer wasn't stalking the rocky shores of the Pacific Northwest. Although *The Mystery of the Floating Feet* would make a great title, it probably won't become a Netflix original documentary— especially once producers discover that their footage would mostly feature crabs dragging pig entrails across the ocean floor, rather than lingering shots of a serial killer's high school yearbook photo.

That's the difference between armchair *CSI* fans and actual forensic scientists: A scientist wants to know the right answer, even if it's mundane. But if you think about it, it's actually pretty exciting that nature hands us clues to what would otherwise probably remain cold cases. Even years later, a missing person might be found, his or her death investigated, all because of a peculiar combination of foot physiology, scavenger behavior, and footwear technology.

Sometimes, such unexpected clues lead us places we never thought we'd go, if only we are willing, patient, and brave enough to follow them. And sometimes they do it wearing sneakers.

THAT'S

DISGUSTING

A BUGGY BUFFET

Why you should be eating insects, but probably aren't

The "Evening of Insect Cuisine" runs $50 a head—$60 if you add two alcoholic beverages. The event is the apex of the 2018 Eating Insects Conference, a three-day symposium where scientists and the edible insects industry gather to share research and eat lots of bugs. Not surprisingly, there's a full house.

An eager crowd of more than 50 diners awaits a dinner party with a twist: cocktail shrimp covered in ants, guacamole with *gusanos* (worms), and chocolate mousse with Japanese wasps perched atop. The chef announces that dinner is running a bit behind, so everyone mills about the banquet room of the Georgia botanical garden, nursing cans of Creature Comfort beer.

Finally, the first dish appears: popcorn with *chapulines,* or Mexican grasshoppers. Servers set trays of plastic cups on a buffet table, and everyone rushes over to grab one. The crickets are smaller than I had imagined, dark brown and glossy, with nary a leg or antenna to be seen. They look like candied nuts, but with tiny faces. I chomp on a couple pieces of popcorn, preparing myself, then slip a grasshopper—the first insect I've ever eaten—into my mouth. The popcorn was the perfect primer. The grasshopper has a nearly identical consistency: crispy and firm, but not crunchy. It tastes mostly like its salty seasoning, and a little toasty.

I'm attending this dinner to challenge myself, having avoided eating insects for years despite covering the "gross" beat. The experience was made easier by the steadying company of my husband, Jay, as well as my dear friend and fellow science writer Susan Milius of *Science News*. (Milius is a longtime vegetarian who agreed to eat animals just this once—but only the six-legged variety.)

The whole premise of this conference is that we should all be eating more insects, because they're nutritious, environmentally sustainable, and—as this dinner is meant to demonstrate—delicious. But one bugaboo hovers over the meeting: the disgust factor. For the five billion or so people on Earth who didn't grow up doing so, the idea of eating an animal that wears its skeleton on the outside is just gross.

Disgust is a tricky quality. We all feel it, but not quite to the same degree or in the same circumstances. The emotion has puzzled generations of scholars who have tried to understand whether the reaction is innate or learned, nature or nurture. And as with much of biology, the answer has turned out to be a little of both.

Some things are disgusting to pretty much everyone over the age of two: feces, parasites, oozing wounds. We even make the same facial expression when we feel disgust; anywhere in the world, you can recognize grossed-out people by their wrinkled noses and the downturned sides of their mouths. When something's *really* disgusting, we instinctively expel it and close off our orifices, squinching our eyes shut and pushing our tongues out of our mouths.

But other disgusting phenomena are particular to certain cultures—especially when it comes to food. Every country in the world has some partially rotted sustenance that it loves—cheese aged with fungi, for instance, or anything fermented. In Iceland, decomposing shark meat called *hakarl* is celebrated as a national dish, while the rest of the world runs away screaming.

In recent years, researchers including psychologist Paul Rozin and public health scientist Valerie Curtis have sought to explain what's disgusting and why. To do that, you first need to be able to measure the intensity of disgust, for which various scales have been formulated. Next, you need to figure out what disgusts people and whether they're *really* as grossed out as they say they are. That's what Rozin set out to discover in 1999, in experiments involving 28 disgusting tasks.

Things started off easy: Student volunteers were asked if they were willing to look at, touch with a fingertip, and finally touch to their lips a Fritos corn chip. But the scale ratcheted up quickly with the next test: a dead, sterilized cockroach. Participants weren't allowed to say they were willing in principle, either; if willing, they had to do it. From there, the experiment tumbled downhill fast, with experimenters hauling out everything from a 10-pound bag of Purina Dog Chow (7 percent of students ate a piece) to an unused tampon (31 percent put the tip in their mouth) to an authentic Nazi officer's hat bearing a swastika medallion (44 percent put it on their head).

As for eating insects, more than twice as many people were willing to stick a straight pin in the eye of an actual dead pig's head (21 percent) than would touch a live mealworm to their lips (9 percent). That surely says something about the challenge the edible-insects industry faces.

From an evolutionary perspective, the roots of disgust stem from an aversion to things that can make us sick. So not only are bodily fluids considered disgusting, but so are the insects that flock to waste and decay, such as flies and roaches. The same is true of animals that are themselves parasites, like lice.

From these practical origins, human beings have expanded the concept, so that not only can sights, sounds, and smells be disgusting, but so can anything we consider morally repugnant. Along with vomit and stickiness, Dutch women listed politicians among things they found disgusting in one survey. So far as we know, human beings are the only species that has elaborated on disgust to such an extent—which explains why we're also the only species with a concept of manners, which serve largely to protect one another from being disgusted. Our manners, in turn, allow us to live in large, complex societies and mostly get along with one another.

In essence, disgust helps make us human.

As it turns out, my own level of disgust for eating an insect (that is, cautiously willing, as long as it's not squishy) is not unusual for an American, and generally predictable for my demographic. Research by the edible-insects industry pegs the typical Western bug-eater (or at least those willing to give it a go) as educated urbanites, 20 to 40 years old, and concerned about the environment.

If I were a man, maybe I'd be a little more enthusiastic. Research shows that men are more willing than women to try a six-legged snack, which is consistent with disgust in general: overall, men rate most things as less disgusting than women do. When Rozin surveyed more than 500 people from the United States and India about their willingness to eat various insects, he found little difference between Americans and Indians, but a significant gender gap. On average, women in both countries said they'd eat a whole, sterilized insect only if their survival depended on it. Men's endorsement wasn't exactly ringing, but averaged out as "maybe."

At the Eating Insects meeting, though, no one is making a disgust face, and men and women toss back insects with equal abandon. Each time servers appear bearing trays of small plates, diners rush to the buffet table to snatch them up. At one point I raise my eyebrows and audibly gasp as a woman dumps a pile of grasshoppers onto her rice dish, walking off with nearly a quarter of the available insects. The nerve!

More choices appear soon, showing off the range of insects now available on the U.S. market for your next soiree: gusano (the worms found at the bottom of a bottle of mescal), weaver ants, termites, black ants. I sprinkle some on a brown paper napkin and take them to a table for identification by edible-insect seller Bill Broadbent of Entosense, Inc. Then, I can savor each species individually.

As it turns out, the lime flavor of the ants, from their formic acid, is quite nice. This is followed by termites (small, nutty) and weaver ants, which have a dry flavor that my husband likens to cardboard. (If there's an "essence de bug," that dry flavor seems to be it.)

The good news is that all insects on the menu are mercifully squish free. "Most of them are completely dried," says Broadbent, describing not just the dinner but the edible-insects market in general. Insects imported from Asia or elsewhere into the United States, he explains, must be preserved to get past the U.S. Department of Agriculture—and for insects, that often means drying them. It's also easier on the American palate, where soft or slimy foods are considered more disgusting than crunchy or crispy ones.

Next, the silkworm pupae dish comes out: a large chickpea fritter covered in a curry sauce with bits of yellow pepper. I extricate a brown pupa covered in curry sauce and examine it closely—a mistake, because it doesn't look better outside the fritter. It's at least an inch long and segmented like an overgrown pill bug.

My pupa is getting cold while I wage an inner battle. But I'm here for a challenge, so in it goes. Maybe because the texture is a bit soft and mealy, or because it's more clearly an insect, but this is the only dish at the meal I don't like enough to finish. When Chef Joseph Yoon stops by to ask how we enjoyed his creations, I find myself leaning to one side to block his view of my half-eaten curry.

By this time, I've nibbled enough bugs that I'm trying to surreptitiously pick flakes of exoskeleton from between my teeth when Marianne Shockley, the meeting organizer, plops down next to me to chat. With a mop of curly brown hair and an infectious laugh, it's easy to see why she's the University of Georgia entomology department's secret weapon for public outreach. If *she* loves bugs, *you* want to love bugs too.

Shockley says she eats insects maybe four or five times a week, and makes a full meal featuring them at least weekly. A little cricket powder in the smoothies for her kids and she "feels like a good mom," she says. I find my head bobbing in agreement, as though I was about to go whip up a cricket smoothie myself.

The next evening, my friend Susan and I compare notes; I offer her some black soldier fly larvae and she picks one from the pile, examines it closely, and then delicately bites it in half. She chews, thoughtfully. "It's not an unpleasant experience," she says. And that sums up our feelings about eating insects in general. We can't decide if we're enthusiastic about it, but we can say that it was, on the whole, not disgusting.

By the end of the conference, I at least *want to want* to eat bugs. It seems like the right thing to do; after all, many well-meaning people are working hard to make them safe and sustainable. Scientists are testing environmental impacts, measuring nutrients, and even considering the effect of insect-eating on our gut bacteria. Growers and sellers are looking for new ways to make insects tasty and raise them organically. One woman even spoke about her organization's efforts to help orphanages in the Democratic Republic of the Congo farm their own palm weevil larvae; nearly half of that country's children are so malnourished that their growth has been stunted, and the weevils are already a popular food there.

But it's hard to suddenly see insects in a new light. No matter how healthy they are for us and the environment, Westerners are unlikely to embrace eating them just because it's logical. Somehow, insects have to transform in our eyes, shedding their exoskeleton of grotesqueness. (Lobsters made the leap, so maybe it's possible.)

If so, I would submit that mainstream bug-eating will probably catch on via the kind of gourmet experience that Chef Yoon delivers, rather than an indistinguishable mash in a protein bar, which is mostly what's on offer now. It's the lure of the exotic—for example, the promise of savory chapulines coated in spicy chili or the crunch of termites said to taste like crispy bacon bits—that will push our curiosity over the edge.

After all, disgust may be hard to overcome, but it's not immutable. As we'll see in the coming stories, we're often most fascinated by the very things that disgust us. I wouldn't have thought I'd enjoy learning about sewer clogs, for instance—but writing about the mammoth fatbergs of grease and waste that grow under cities made me want to go see one for myself.

Likewise, don't be surprised if you find yourself enjoying reading about body fluids and things that stink in this chapter. (Who knows, you might find yourself with some new cocktail party conversation starters.) First, we'll test your tolerance by delving into the most disgusting edible insect I've come across: maggots. They're the next big thing for feeding the world—although the first time you taste food made using maggots, you probably won't even know it. And that's the biggest challenge: taking the first bite.

ON THE MAGGOT FARM

How fly larvae are squirming into our food chain

Apart from driving an enormous Ford F-250 pickup truck, Jeff Tomberlin doesn't particularly look like a farmer. He wears stylish black-framed glasses above a shock of facial hair that's just shy of hipsterish. Then again, Tomberlin farms maggots, and it's hard to say what maggot farmers should look like. (Based on my limited sampling, they wear cornflower blue sweater vests and drive large trucks while listening to hard rock from the 1980s and '90s.)

Tomberlin radiates an upbeat confidence, a can-do spirit that probably comes in handy when explaining to strangers that he has spent much of his career studying the squirming babies of flies. His interest began with the kind of maggot that tends to appear on dead people; this led to his becoming one of North America's 20 or so certified forensic entomologists. (This special class of scientists specializes in studying insects on cadavers; maggots are some of their main tools.) Today, in addition to helping solve murders—including providing key testimony in the exoneration of a Las Vegas woman for a murder she didn't commit, based on the *lack* of maggots on the victim—he runs a laboratory focused on the biology of the black soldier fly. He also directs a maggot-farming company, EVO Conversion Systems, which advises companies on how to farm black soldier fly larvae by feeding them organic waste. This, in turn, yields a rich feed for chickens, cows, fish, and other livestock.

Tomberlin makes maggot farming look very normal, and supremely practical. It appeals to his love of efficiency, which is stronger than most people's. (His email inbox contains four messages—total.) I visited him in College Station, Texas, a few months after the Eating Insects Conference in Georgia. It's clear he's a celebrity of sorts among those who farm fly larvae as food. And his following is growing; black soldier fly larvae are a hot topic among insect growers, with maggot farms springing up across the United States and around the world to meet the growing demand.

The black soldier fly itself is a long-bodied black creature, rather waspy-looking, with thick wiggling antennae and wings tinged in a lovely iridescent blue. It looks and acts almost nothing like a housefly. "You'll never see one in your house," Tomberlin tells me as we peer into a cage filled with tens of thousands of them that are busy mating and laying eggs in the wavy bits of cut cardboard. They seem to have little interest in, or fear of, people.

What does interest them is anything that's decomposing. In its two weeks or so as an adult, a black soldier fly's primary mission is to mate—and if it's female, to lay its eggs on or near something rotten, so that its maggot babies will have plenty of decayed food or manure to feast on. This is the basic idea behind sustainable maggot farming: Feed the maggots low-value food waste (which humans produce at the astonishing rate of 1.3 billion tons a year), so that they themselves can be transformed into a cheap and environmentally sustainable supply. And that, my friends, is the magic of maggots.

In China, the larvae are already being grown by the ton, and Tomberlin's consulting service is in high demand. Chinese maggot-farming operations produce feed for all manner of aquatic animals, from farmed fish to frogs and eels, and that livestock in turn helps feed the nation's 1.4 billion people.

In fact, if you can get past the gross factor, eating the animals that eat fly larvae makes a lot of sense. We use about a third of American cropland to produce feed for livestock, with a heavy toll on the environment. Insects require less land and water to grow, so they could be a more sustainable alternative.

And what about feeding black soldier fly larvae to people directly? After all, wouldn't that be even more efficient than feeding them to animals and then eating the animals? It would, Tomberlin says. "But that's a bigger challenge, convincing people that eating insects is natural, and it's good for you, and it's good for the environment." And of course, there's the disgust factor. The word "maggot" seems like a barrier, I observe. Come to think of it, "larvae" isn't a whole lot better. "The word 'maggot' has a negative connotation to it—so that's the hard part," Tomberlin confesses.

Still, if crickets are making their way into snack bars, Tomberlin thinks black soldier fly larvae could too. "I actually think they're better than crickets, but I'm biased," he says. He describes them as creamy and buttery when eaten fresh, because of their high fat content. "I sort of look at it thisaway," he says. "It's like a wine tasting—there are some people who are skilled enough to taste the difference between a high-grade and a low-grade wine. Maybe the same will happen with insects."

The morning after my visit with Tomberlin, I arrive at the aptly named FLIES Facility (short for the Forensic Laboratory for Investigative Entomological Sciences). It's a cluster of small white buildings and a greenhouse filled with fly cages, tucked in the nether regions of the Texas A&M University campus. Compared with the massive six-story building housing the entomology department located several minutes' walk away, the fly facility looks like a place where gardeners might keep equipment for maintaining pristine lawns elsewhere on campus.

My tour guides to the maggot-growing operation are Jonathan Cammack, a postdoctoral fellow, and graduate student Chelsea Miranda. As they lead me into the fly lab, I immediately figure out why the other entomologists have required Tomberlin to keep his maggots a quarter mile away from them. The maggot lab stinks.

"What is that particular aroma?" I ask Miranda as we walk into a room where a wall is lined with large bags labeled "FEED."

"Decomp," she says. As in decomposition. Just about any organic material, whether animal or vegetable, produces a recognizable mélange of chemicals as bacteria break it down. Some have names like putrescine and cadaverine, which smell like they sound. Another, a subject of study in the FLIES lab, is indole, a breakdown product of the amino acid tryptophan. At low levels, indole contributes to the heady scents of jasmine and orange blossom; at high concentrations, you'd recognize it as one of the primary smells in poop—and it attracts flies like crazy.

What does the indole/putrescine/cadaverine mixture in the fly lab smell like? The best I can describe it is sweet-and-sour pork left in a barn to rot. "I'm getting more of the sour today," Miranda says when I bring up the comparison. Other days, it's sweeter.

Cammack leads me into a small, toasty incubation room where the maggots live. They're housed in large plastic trays stacked floor to ceiling on metal shelves, where they're currently chowing down on decomposing brewers' waste, a mix of barley and other grains left over after brewing beer that contributes to the sour aroma. Brewers call these leftovers spent grain. On its own, spent grain is mostly fiber. But when maggots eat it, they work their biochemical magic, converting it into fats and proteins in their growing bodies.

As it turns out, black soldier fly larvae have as much protein as chicken, according to Liz Koutsos, president and CEO of EnviroFlight, a company that turns maggots into feed and fertilizer. EnviroFlight is already producing feed for poultry and salmon, and is working on larvae-based pet food, too.

So little by little, maggots are worming their way into our food supply. For the most part, they're still raised using traditional feed: grains and such. But the ultimate in maggot farming—you might say the holy grail of maggotry—would be to raise them on waste:

manure or other organic material that would otherwise be thrown out. I can see how it appeals to Jeff Tomberlin's highly tuned sense of efficiency; in this process, nothing goes (ahem) to waste.

Our goal in the fly lab is to collect 1,200 maggots from the trays, or about 100 grams. Then Cammack will freeze them and send them to collaborators who will analyze their composition: levels of fats, proteins, and other substances that will determine the optimum amounts of maggot to mix into other feeds.

When I ask why not just feed animals 100 percent maggots, Cammack tells me the larvae are actually too packed with proteins and fat to be the sole feed source for most livestock. "They're basically like eating a Snickers bar," Cammack says.

Cammack slides out a white tray, and I look inside. I expect something like wiggling rice grains, but all I see is damp, nonwiggling brown stuff. It looks like the leftovers you might find under a bird feeder after a hard rain. He assures me that there are up to 10,000 or 12,000 maggots in each tray, but they tend to burrow down into their food, avoiding light and seeking out damp, dark spots.

Miranda uses this tendency to her advantage to separate the maggots from the grain. After donning bright blue gloves, she uses her hands to scrape a thin layer from the top. After a swipe, the entire surface of the grain suddenly ripples with thousands of maggots frantically digging themselves deeper. Miranda patiently scrapes a thin layer at a time from the surface, driving the maggots to the bottom of the pan, where she can collect them en masse and divvy them between the three of us.

One by one, we start quietly picking maggots from our respective piles and dropping them into clear plastic containers that look suspiciously like the ones hot and sour soup is delivered in. And in fact, they are the same kind, purchased from a restaurant supply store. This adds to my growing queasiness from the smell, which has reached the level of a Dumpster full of rotting sweet-and-sour pork in a barn.

But this experience is far from the grossest thing that these scientists have endured in the name of ensuring a sustainable food supply. "So what do you think about when you're picking larvae?" I ask. "It depends what the substrate is," Miranda says. "If it's manure, sometimes I'm like, why did I sign up for this?" She tells me that two weeks before, she had an existential moment when she plunked a pan of poultry poop down on a lab bench, and it splashed in her face. "That's when I had this little epiphany, like, this actually sucks."

Miranda started off in veterinary science, so she was the natural choice to experiment with raising black soldier fly larvae on livestock manure. She's now grown maggots on poultry, swine, and dairy cattle manure. The chicken shit is definitely the worst, she reports, because she has to climb under cages of laying hens and catch their poop in buckets as it rains down. "And you get down on your knees under the cages, and you're getting it on you. And you have about three and a half feet of room, so you're bent over all the time," she says.

"You're a brave woman," I tell her.

"That's one way to think of it," she says. She's quick to add that she loves her work (apart from the poop in her face), and especially likes the flies—which is funny, she says, because she's afraid of insects. "Wait," I say. "You're an entomologist who's afraid of insects?"

"Anything that bites or stings," she says. (Luckily for her, black soldier flies don't bite.)

In Miranda's blue-gloved hand, a maggot squirms in the light; the head has a pointy bit, which contains its tiny nonthreatening mouth. It's utterly nonscary. And then I realize something surprising: The more you look at maggots, the less disgusting they seem.

As with any other exotic food source, I suppose it's all about familiarity. Sardinians, after all, consider a wheel of *casu marzu* cheese crawling with maggots a delicacy. As for the rest of us, we'll have to ease into eating them—one step on the food chain at a time.

STINKS SO GOOD

The distinct pleasure of smelling something awful, explained

A corpse flower smells like a heady mix of rotten fish, sewage, and dead bodies. It's a stench meant to draw flies—but just as surely, it draws tourists. One blustery Chicago night in 2015, thousands of people lined up at the Chicago Botanic Garden for a whiff of a four-and-a-half-foot-tall corpse flower named Alice.

In fact, the demand to see and smell this specimen, known scientifically as titan arum, is so great that botanical gardens now vie to own one. Gardeners lavish them with care, hoping to force more stinky blooms from a plant whose scent is so rare (up to a decade between flowerings) and so fleeting (eight to 12 hours) that visitors are often disappointed to miss peak stench.

But why do people *want* to smell the thing? The reaction is usually the same: anticipation, a tentative sniff, then the classic scrunched-up face of disgust. And yet everyone seems happy to be there.

It turns out this phenomenon has a name: benign masochism. Psychologist Paul Rozin described the effect in a 2013 paper titled "Glad to Be Sad and Other Examples of Benign Masochism." His team found 29 examples of activities that some people enjoyed even though, by all logic, they shouldn't. Many were common pleasures: the fear of a scary movie, burn of a chili pepper, pain of a firm massage. And some were disgusting, like popping pimples or looking at a gross medical exhibit.

The key, Rozin says, is for the experience to be a "safe" threat. "A roller coaster is the best example," he told me. "You are, in fact, fine, and you know it. But your body doesn't, and that's the pleasure." Smelling a corpse flower is exactly the same kind of thrill, he explains. You're smelling something so wretched it feels like it could make you sick, and yet your brain can just gleefully override that message and say, "Everything's fine! It's just a flower!"

Toying with our own defenses in this way is a bit like kids playing war games, says Valerie Curtis, a disgust researcher at the London School of Hygiene and Tropical Medicine. "The 'play' motive leads humans to try out experiences in relative safety, so as to be better equipped to deal with them when they meet them for real."

So by smelling a corpse flower, Curtis posits, we're taking our emotions for a test ride. "We are motivated to find out what a corpse smells like and see how we'd react if we met one," she explains.

And as it happens, our sense of disgust serves a larger purpose. As Curtis reports in her book, *Don't Look, Don't Touch, Don't Eat,* the things we find most universally revolting are those that can make us sick. You know—things like a rotting corpse.

And yet our sense of disgust can be particular. Most people, it seems, are fine with the smell of their own farts—but not someone else's. Disgust tends to protect us from the threat of others, rather than from our own grossness.

Likewise, the same scent compound can elicit different reactions; as perfumers well know, some smells are good only in small doses. Musk, for instance, is a base note of many perfumes but is considered foul in high concentrations. Ditto indole, a molecule found at low levels in white flowers like jasmine, but often described as fecal and even repulsive at high concentrations. You can experience it in a perfume called Charogne,*

---

* I had to try it for myself, so bought a sample vial. At first whiff I found Charogne overwhelmingly floral—and sure enough, it contains lily and jasmine as well as ambrette, a plant with a musky scent. After wearing it for an hour, I found the smell cloying but wouldn't describe it as carrion per se—more like a wilting funeral arrangement.

which translates to "carrion." On the market since 2007, it has some fans.

To my knowledge, no one has yet tried to create a corpse flower perfume. But the possibility remains, because, after all, its scent is just another mix of compounds: indole, sweet-smelling benzyl alcohol, and trimethylamine (also found in rotting fish). This magical concoction could be achieved using headspace technology, pioneered by fragrance chemist Roman Kaiser in the 1970s. The process captures a flower's fragrance by enclosing the blossom in a glass vial to obtain the molecular mix that wafts from it, then re-creates the blend of scent chemicals. So eau de corpse flower could be on the horizon—if someone can find a large enough headspace vial!

My own favorite "love to hate" smell stems from my childhood in the 1980s. At a time when I loved Strawberry Shortcake dolls and scratch-and-sniff stickers, the boys in my class were playing with He-Man dolls (excuse me, action figures). Among the coolest, and grossest, was a character called Stinkor. He was black and white like a skunk, and his sole superpower was to reek so badly that his enemies would flee, gagging.

To give Stinkor his signature stink, Mattel added patchouli oil to the plastic he was molded from, ensuring that his scent couldn't fade (unlike those of my Strawberry Shortcake dolls). The smell was one with Stinkor—and of course, children loved him. Writing on the perfume website Basenotes about the Stinkor figure that she and her brother adored, Liz Upton beautifully captured his unique appeal: "Something odd was going on here. Stinkor smelled dreadful, but his musky tang was strangely addictive," she confesses. "Of course, we scratched and sniffed again, and again, and again, until poor old Stinkor wore a hole in his chest."

If you're the kind of benign masochist who wants to smell Stinkor for yourself, you can pay $125 or more for a rereleased collectors' edition—or you can just find an old one on eBay. Because believe it or not, the original Stinkor dolls still stink 30 years later. And people still buy them.

I can relate. After my research into "carrion" perfume, I got curious about its maker's other unusual offerings. Some sounded intriguing in a horrible way—for example, Jasmin et Cigarette, described by reviewers as reminiscent of a wealthy chain-smoker, and Malaise of the 1970s (which, sure enough, contains patchouli). Another scent, called I Am Trash, incorporates the waste created in the making of other perfumes; in a promotional video, earthworms twist and squirm on a bed of moldy orange peels. Weird—and yet I found myself debating a $39 sample collection.

The experience left me mulling another reason we might sometimes long for stinky or otherwise unpleasant smells: nostalgia. For some people, Stinkor smells like childhood play; for others, cigarettes smell like grandma. These emotional connections run deep in our brains; smell is often said to be the sense most closely linked to memory. My own favorite perfumes are heavy on the scent of orange blossoms (which also hold a hint of indole); one sniff and I'm instantly transported back to a grove in Florida, where I once sat in the crook of a tree eating a fresh orange as its juice dripped down my wrists.

Scientists call this the Proust effect. In the celebrated author's novel *In Search of Lost Time,* the smell of freshly baked madeleines steeped in tea triggers a flood of vivid memories of the narrator's childhood. Neuroscientists have suggested that this effect is the result of activity in the brain's olfactory bulb, which processes odors; as it happens, the olfactory bulb is directly connected to the hippocampus and the amygdala, parts of the brain involved in memory and emotion. Some experiments have supported this hypothesis; a 2012 study by Utrecht University researchers determined that smells triggered more intense and detailed memories than sounds did, even for unpleasant recollections.

So perhaps occasionally, we might learn to associate a bad smell (as long as it's not *too* disgusting) with a happy experience, like a trip to the botanical garden to smell a corpse flower. After all, a picture may be worth a thousand words—but one whiff can hold a lifetime of memories.

PASS THE SEMEN

How blowflies can impact crime scenes

People sometimes ask me, "What's the grossest thing you've ever written about?" My answer is usually that it depends what you, as a unique human being, find especially gross. Because we're all not equally disgusted by the same things, it follows that you might be most horrified by eye worms, while others are more repulsed by, say, head transplants.

But if you ask about my favorite gross scientific experiment. . . well, I do have an opinion on that. My nod goes to forensic expert Annalisa Durdle's experiments feeding bodily fluids to blowflies. Flies' taste in people juices turns out to be not only a delightfully gross topic, but also surprisingly important for criminal justice.

Blowflies, in case you're not familiar with them, are insects of death. These big shiny insects have an amazing ability to appear, seemingly out of nowhere, within moments of blood being spilled, or at the slightest whiff of decay. As a result, hordes of them are often found buzzing around a gory crime scene. That got Durdle wondering: With all those flies doing what flies do—flying around and pooping on stuff—could they be contaminating crime scenes?

"Interestingly, fly poo can also look very similar to blood spatter," says Durdle, a lecturer in forensic science at Australia's Deakin University. Given this, she explains, entomologists occasionally get called in to identify exactly which of the two substances crime scene investigators are contemplating.

Even for an expert, it can be hard to tell the two apart. In fact, I saw a presentation at a forensic science conference in which a scientist had resorted to scanning electron microscopy to tell blood from fly poop, because it's so often impossible to distinguish them based on color or shape of the droplets. Confusing them could throw off an analysis of blood spatter patterns, making it look as if blood was spilled during an attack when really, a fly just pooped there.

Another reason it's important to tell the difference between these substances is that speckles of fly poop could also contain the DNA of the person whose fluids they snacked on—and that person may or may not have been involved in the crime. "It turns out that you can get full human DNA profiles from a single piece of fly poo," Durdle observes. "I tend to refer to poo rather than vomit," she clarifies, "because in my experience flies tend to eat their vomit, and most of what you have left is poo—although they do eat that too!"

Clearly, blowflies are gross. But could they really incriminate someone by eating body fluids? To find out, Durdle needed to know what blowflies would be likely to eat at a crime scene. So she experimented, offering Australian sheep blowflies a crime scene buffet with body fluids collected from volunteers—blood, saliva, and semen—plus other appealingly high-protein and high-carbohydrate snacks that flies might find in a victim's home: pet food, canned tuna, and even honey.

Assembling the buffet involved a bit of delicate collection. Volunteers had blood drawn and spat into a tube to provide saliva samples, and "semen donations were self-generated by masturbation into plastic specimen containers and stored at 15°C until required," according to the research paper, which was published in the *Journal of Forensic Sciences* in 2016. Durdle told me she was eternally grateful to all the volunteers, who remain anonymous. "Particularly the one who had to hide his sample pot in the frozen peas because he knew his stepdad didn't like peas and would never find it there," she says.

In any case, the offerings were assembled and placed in little dishes; next, the hungry flies were released. Durdle's team recorded each group of flies (including males and females of different ages) on video for at least six hours to observe their feeding behaviors and preferences for the various menu items.

They say you catch more flies with honey. But in this case, they're wrong. What you catch more flies with, it turns out, is semen.

"It's the crack cocaine of the fly world," Durdle says. "They gorge on it; it makes them drunk. They stumble around, partly paralyzed—I've even seen one fly give up hope of cleaning itself properly and sit down on its bum!" she says. "Then they gorge some more and then it kills them. But they die happy!"

The flies would be perfectly satisfied to eat pet food, blood, or saliva if offered individually—but if semen was an option, most preferred it above all else. As for why semen was such a hit, it may be that flies are particularly drawn to its smell. Flies are attracted to odors associated with decay, which include compounds like cadaverine and putrescine (known chemically as polyamines)—and sperm contains related polyamines called spermine and spermidine. A 2016 study published in the journal *PLOS Biology* found that flies are not only highly attracted to polyamines, but also need them nutritionally. When female flies ate food enriched in polyamines, they laid three times more eggs.

Another thing semen is full of: DNA. So if a fly stuffs itself with this substance, it's also stuffing itself full of the semen donor's DNA, which the fly could in theory expel someplace else.

To see if that ever happens, Durdle also tested flies' poop after they'd had their fill of various offerings from her buffet. "If the flies had fed on semen or a combination with semen in it, then you got a full human DNA profile almost every time. With blood, it was maybe a third of the time, and with saliva, never."

Durdle didn't stop there, as one might expect from a scientist willing to look for semen leftovers in insect poop. Why not also see whether flies prefer dry or wet food, like a finicky cat?

"It was also interesting to find the flies generally preferred dry blood or semen to wet blood or semen," Durdle says. "This could be important, because it means flies could continue to cause problems at a scene long after the biological material had dried."

In a 2018 study in the *Journal of Forensic Sciences*, Durdle's team released flies into a house[*] where blood, sugar, and water were the only food sources, then watched to see where the flies would poop. They determined that crime scene investigators should be wary of blood spots found near food sources, particularly in low places—because they could well be fly poop.

How big a deal is this? "You really need to look at the probabilities, including the chance that a fly might feed on some poor guy's semen after he's had some innocent quiet time to himself, and then fly into a crime scene and poo, potentially incriminating him," Durdle observes.

Chillingly, the scenario doesn't sound entirely implausible—and not just in the case of overly enthusiastic teenage boys. A man who innocently left some sperm behind after a sexual encounter might have cause for concern about where his DNA might show up later. There's also the chance, Durdle says, that a forensic investigator could sample fly poop thinking it's blood spatter, and discover DNA that's not from the victim.

This business with fly poop is just one part of a bigger problem that's starting to crop up in forensic DNA analysis. As technology used to detect DNA has become more sensitive, there's greater risk of finding tiny bits of DNA that got there accidentally, as forensic scientist Cynthia Cale observed in the journal *Nature* in 2015. Moreover, it doesn't have to be a fly carrying DNA from place to place; even a handshake can do it.

....

[*] This immediately made me wonder, *Whose house* was it? It was not Durdle's. It was a house that no one lived in; researchers cleaned it of anything flies might eat and covered all the surfaces with butcher paper, then sealed the flies in with masking tape around the doors.

In fact, Cale demonstrated that one person can transfer another's DNA to a knife handle after just two minutes of holding hands. If that sounds like a lot of contact, it did to Cale too. So she also experimented with people touching for 60, 30, and 10 seconds. The verdict? DNA transfer was detectable in most cases. Even with as little as 10 seconds of contact, she detected what she calls a "contributor inversion," meaning that the majority of the DNA on the knife came from a person who had never actually touched it.

And as with flies inadvertently carrying DNA to a crime scene, this could be a real problem. "I think the biggest impact might be when a defense lawyer uses it to raise doubt in the mind of a jury," Durdle says.

As it happens, there's been at least one case of false accusations based on transferred DNA. In 2013, Lukis Anderson was arrested in California for murder and held for four months because his DNA was found under a victim's fingernails. That's pretty incriminating, usually. But it turned out that Anderson had an airtight alibi: He was in the hospital exactly when the murder happened, detoxing from a drinking spree. He'd been so drunk, in fact, that at first even Anderson himself wasn't entirely sure whether he'd done it.

Eventually, a police officer connected the dots: Three hours after paramedics hauled Anderson to the hospital, the same team was called to the scene of the murder and handled the victim's body. Somehow, they left a little of Anderson's DNA behind.

For now, no one really knows how often mix-ups like this might happen. But it's going to be important to figure that out, says Cliff Spiegelman, a statistician at Texas A&M who specializes in forensic science. "There are very few studies, and the right answer is they just don't know," Spiegelman says.

Nor is it clear whether anyone has ever been wrongly convicted based on DNA collected from fly poop; it's just clear that it's physically possible, thanks to studies like Durdle's. But the news is not all bad; in fact, it's even conceivable that a fly could help deliver criminal justice. If, for example, it eats bodily fluids from a crime scene, flies into another room, and poops there, that poop might preserve a DNA sample and foil a perpetrator's attempts to clean up.

What it all comes down to is that forensic investigators need to consider all potential sources of DNA—and be extremely careful not to accidentally transfer DNA from one object to another. As for the rest of us, it's a little disconcerting to think that our genetic material can be spread so easily. All we can do is be careful whose hand we shake—and watch out for flies.

SNIFFING OUT SICKNESS

Harnessing the smell of disease

I am sick, and I don't smell right. I don't mean that my nose isn't working—though this cold has me stuffed up. Instead, my own body odor seems somehow different: sour and unfamiliar.

I'm far from the first person to notice this nasty side effect. Scientists have found that dozens of illnesses have a particular, identifiable smell: Diabetes can make your urine smell like rotten apples, and typhoid turns body odor into the smell of baked bread. Worse, it's said that yellow fever makes your skin smell like a butcher's shop, if you can imagine that.

It's curious, but not merely a curiosity; our noses and brains are attuned to these smells, which in turn signal our finely tuned sense of disgust to feel grossed out and thus help us avoid something that could make us sick.

We might even be able to harness this "sickness sensing" ability. Scientists think that if we could identify the particular chemicals that make up sick smells, we might sniff out diseases that are otherwise difficult to detect early, like cancers or brain injuries. Some people already have a surprising ability as disease detectors (more on this later). If we could replicate that skill, perhaps annual physical exams of the future could include a sniff test for early stage Parkinson's and other diseases.

In fact, anyone with working olfactory senses could probably learn to recognize various "sick smells." Humans are very good at detecting illness, says Valerie Curtis, a public health expert at the London School of Hygiene and Tropical Medicine who studies disgust.

Curtis says, "Signs of sickness are some of the things people find most disgusting"—for example, mucus, vomit, or pus. Because disgust is our way of avoiding things that could harm us, Curtis continues, "it simply makes good evolutionary sense that we use our noses to notice illness."

But why would sick people smell differently in the first place? The key is that our bodies are constantly launching volatile substances into the air. They're carried in our breath and ooze from every pore, and can vary depending on age, diet, and whether an illness has thrown off some cog in our metabolic machinery. Microbes living in our guts and on our skin also contribute to our signature scent by breaking down our bodies' metabolic by-products into smellier ones.

So let's face it: Basically, you're a walking factory of smells. And if you start paying attention to them, you might notice when something's off.

Parkinson's is notoriously tricky to diagnose; by the time most people learn they have it, they've already lost half the dopamine-producing brain cells that the disease attacks. But about 12 years before her husband, Les, was diagnosed in 1995, Joy Milne noticed that he smelled odd. Les had a "sort of woody, musky odor," Milne told the *Telegraph*, a British newspaper. Years later, in a roomful of Parkinson's patients, she realized the smell wasn't unique to her husband; they all smelled that way.

She mentioned it to a Parkinson's researcher in Edinburgh named Tilo Kunath, who mentioned it to his colleague, analytical chemist Perdita Barran. They decided the well-meaning Mrs. Milne may have just noticed the characteristic smell of old people. "We talked ourselves out of it," Barran says.

That could have been the end of it. But another biochemist encouraged the pair to track down Milne and try a blind T-shirt test: She sniffed six sweaty tees from people diagnosed with Parkinson's, and six from healthy controls. Milne correctly identified which six had Parkinson's, but she also tagged one of the control subjects as having the disease. Despite that error, Barran was intrigued—and all the more so eight months later, when the same supposedly healthy control subject Milne had identified was diagnosed with Parkinson's. Had she really detected the disease before doctors did?

The T-shirt test was intriguing, but we have to take it with a scientific grain of salt. After all, it involved a small number of people, and people might share an odor for lots of reasons. In one notorious dead end, researchers were once convinced a smell was linked with schizophrenia; a particular compound called TMHA—said to smell like goats—was identified and described in the journal *Science*. There was hope this chemical might even be the cause of schizophrenia, which would open up new avenues for treatment. But in years of follow up testing, the results couldn't be repeated. The TMHA "schizotoxin" eventually went the way of cold fusion.*

Barran is now a professor at the Manchester Institute of Biotechnology, where she's applying the painstaking methods of chemistry to determine whether the Parkinson's smell is the real deal. She and her colleagues aim to develop a smell test for Parkinson's—one more rigorous and more practical than having Mrs. Milne over to smell all our T-shirts.

First, the team is working to chemically identify the molecules involved, which is harder than it looks on *CSI*. Of the thousands of known volatile compounds, many are not well characterized, or data on them exist only within the fragrance industry.

With funding from Parkinson's UK and the Michael J. Fox Foundation, Barran's team collected more than 800 samples of sebum, an oily substance secreted by skin, swabbed from

....

* In 1989, Stanley Pons and Martin Fleischmann announced with great fanfare that they had smashed atoms together at room temperature and produced energy. The breakthrough promised a virtually limitless supply of energy. Unfortunately, no one could replicate their findings, and cold fusion became the laughingstock of physicists worldwide.

the backs of volunteers. In 2019, they tested sebum from 64 volunteers and reported finding three molecules that are elevated in people with Parkinson's disease (eicosane, hippuric acid, and octadecanal) and one that's lower (perillic aldehyde). Then they blended these molecules to create a chemical fingerprint for the disease. When they presented their signature scent to Joy Milne, she confirmed that it smells like Parkinson's disease to her, too.

Next, the team not only has to confirm that these particular molecules are reliably elevated in Parkinson's patients, but also figure out whether they can detect the smell before symptoms of Parkinson's appear. Ideally, they would also learn how Parkinson's triggers the body to produce the molecules.

Barran says she's up for the challenge—even though her own sense of smell was damaged in an accident, and she can't smell the Parkinson's odor herself.

"Joy [Milne] has an extremely good sense of smell," Barran says, "but she isn't the only person who can smell it. What's special is how persistent she was in her conviction that it was something that could be used [to conquer a disease.]"

This brings us back to the question of what you and I can actually smell. Although dogs have been the most lauded in this regard and have been tapped to sniff out cancer, research suggests that humans are just as good at detecting many odors.

Judging by the number of neurons in our brains' olfactory bulbs, people may be better smellers than rats and mice, and fall about in the middle of the pack among mammals. Perhaps the biggest barrier to our abilities is that we don't pay enough attention to smells, and we lack a sophisticated language to describe them.

"We're less able to rationalize smell," says Curtis. She recalls using a soap she had brought home from India: "The idea of 'India' popped into my head long before I realized it was the smell."

Likewise, we may not realize when we're smelling a change in our own or a loved one's health, simply because we're not accustomed to paying such attention to scents.

There are hints, though, that we may be decent illness detectors if we pay attention to the task. In a small double-blind study published in the *Proceedings of the National Academy of Sciences* in 2017, participants could identify sick versus healthy people based on body odor and photographs just a few hours after some of the people's immune systems were triggered by a toxin that mimicked infection.

So, although we don't yet have a Breathalyzer for disease, we might do well to follow our noses.

SEWER MONSTERS

Inside the growing scourge of fatbergs

First, someone might pour molten turkey fat down a drain. A few blocks away, someone else might flush a wet wipe down a toilet. The two scraps find each other in a dank sewer pipe, and eventually, more fat, oil, and grease congeal onto the mess and build up into giant stinking globs. Voilà! A baby fatberg is born.

Like the floating ice that sank the *Titanic,* fatbergs are big, dangerous, and mostly hidden from view. But unlike their floating counterpart, these repellent objects are made of grease and trash that hardens in a sewer system. When they get big enough, fatbergs can clog massive underground pipes entirely, sending raw sewage gushing into streets.

In September 2017, workers in London discovered one of the biggest fatbergs ever seen in the East End neighborhood of Whitechapel. Measuring more than 800 feet long, the monstrosity weighed an estimated 143 tons and was the size of 11 double-decker buses, according to the London utility provider Thames Water. "We reckon it has to be the biggest in British history," a supervisor for Thames Water told the BBC.

Within two years, an even bigger fatberg snapped sewer pipes a meter wide in the London neighborhood of Chelsea. In 2017, Northern Ireland Water excavated "a couple of hundred tonnes" of grease and debris from a fatberg underneath a row of fast-food restaurants in Belfast.

Fatbergs are fast becoming a sewer scourge around the world. London, Belfast, Denver, and Melbourne are just a few of the metropolises where huge versions have been spotted in recent years. In New York City, grease causes 71 percent of sewer backups, according to the city's 2016 State of the Sewers report. The city spent $18 million over five years fighting fatbergs, and its treatment plants rake 50,000 tons of flushable wipes and other crud out of raw sewage every year. Smaller cities aren't immune; Fort Wayne, Indiana, has spent half a million dollars a year cleaning grease out of sewers.

The United States and United Kingdom report the most fatbergs globally, says engineer Thomas Wallace of University College Dublin, who studies the disposal of fat waste. Not only do both nations produce copious fatberg ingredients, but they also possess aging sewer systems ill equipped to deal with the onslaught of fat and trash from growing populations.

Clogs are a problem as old as sewers themselves; the ancient Romans reportedly sent slaves underground to clean their sewers. But the enormous fatbergs of today are brought on by more modern inventions.

The first fatbergs probably started small, as cities and their cooking waste grew with the industrial age. In 1884, Nathaniel Whiting of San Francisco patented the first grease trap to catch "substances which would tend to choke and clog the sewers." His basic design is still used today: Wastewater drains into a box where fat settles out. Eventually, someone has to clean the gunk out and dispose of it.

In the United States, many cities eventually required restaurants and other food sellers to use grease traps and to clean them out; a surprising amount of controversy and intrigue has grown around these caches of fat. In some places, thieves have blowtorched their way into grease traps to steal used cooking oil that can be made into biofuels.

In China, trapped fat is illicitly scooped, cleaned up—though not enough to avoid safety concerns—and sold on the black market as "gutter oil." The Chinese government has banned this substance but struggles to stamp it out. In cheap restaurants and street stalls, your dinner might even be cooked in the stuff; in 2018, the *South China Morning Post* reported that owners and staff of three restaurants had been jailed for cooking hot-pot meals in the goop.

For London, haphazard use of grease traps, combined with narrow sewer pipes, spells disaster. As a result, utility companies have to hire teams of "flushers," people charged with digging out clods of fat by hand. So it's fitting that the term "fatberg" was coined by the people who know them best: the sewer workers of Thames Water.

As fatbergs grow beneath us, scientists are racing to learn more about them. For starters, they recently discovered that most of a fatberg's mass is actually a form of sewer-made soap.

In 2011, Joel Ducoste of North Carolina State University and his team reported that the same process that turns lard into soap, called saponification, can happen in sewer grease if calcium is present. The team even grew miniature fatbergs on calcium-rich concrete in the lab, offering a clue as to how the blobs get so massive.

And in places where fatbergs are on the rise, sewer managers point to wet wipes as a major part of the problem. These premoistened toilet wipes are made for both babies and adults, and although many are advertised as flushable, poorly dissolved wipes are pulled from sewers by the ton. Worse, the tough cloths serve as excellent building blocks for fatbergs, snagging grease and other residue as they flow past.

Tom Curran of University College Dublin received the first Fulbright scholarship awarded to a scientist in the fatberg fight. Curran worked with Ducoste in North Carolina to map potential hot spots and develop sensors that might one day alert cities to 'bergs before they reach pipe-bursting sizes.

Some cities are even looking at fatbergs as fuel. After all, fat is high in calories—and therefore energy. Thames Water has partnered with a renewable-fuel company to dig fatbergs out of sewers and turn them into biodiesel.

However we handle them, fatbergs clearly fascinate us with their sheer grossness. The Museum of London put a chunk of the Whitechapel 'berg on display, and visitors eagerly watched it sweat, change color, and hatch flies. Now you can watch the chunk live on the museum's online FatCam—as of this writing, it had sprouted yellow pustules of mold.

PART THREE

BREAKING

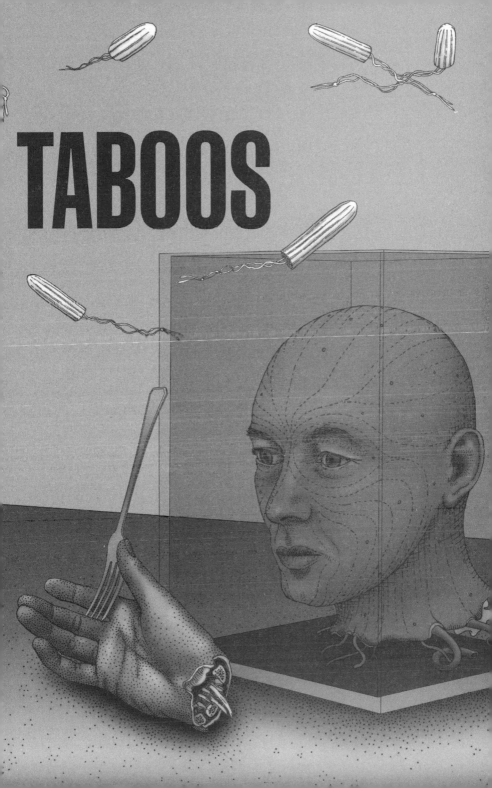

TABOOS

THE ULTIMATE TABOOS

When science intersects with our deepest inhibitions

"Taboo" is a lush, tantalizing word, one that speaks of tempta-
tion and secrets. And for this we can thank the Polynesian
island nation of Tonga, which gave to the English language a
new way to convey an idea beyond "forbidden" or "banned"—
flat, authoritarian terms that lack the spiritual and moral rich-
ness of "taboo."

We can also thank the shaky transcription skills of famed
British explorer Captain James Cook. The main island of Tonga
is called Tongatapu, which Cook apparently heard as Tongata-
boo. Likewise, he misheard the Tongan word for the forbidden,
which was actually *tapu*. Instead, Cook wrote it as "taboo."

Polynesia may have contributed the semantics for the concept,
but every culture shares a notion of the taboo, those acts or ideas
that are morally or spiritually forbidden (or in many cases, banned
from discussion). And just as what's considered disgusting varies
from place to place but generally sticks within certain themes,
what's considered taboo depends where you grow up and tends to
share common features.

Usually, taboos are associated with death and killing, sex and
reproduction (including incest), the dead (including necrophilia),
or food (including cannibalism, which handily combines food
and the dead). Our bodies, and notions about their cleanliness
or purity, also figure prominently.

During his tours of the Pacific, Cook saw how complex taboos can be. When he invited several high-ranking Tongans to dinner aboard his ship, "not one of my guests would sit down or eat a bit of any thing that was there," he wrote. Later, it seemed that any food or drink made using water was *tabu* under some circumstances. Eating with hands that had washed the corpse of a dead chief was also "tabu," requiring that the corpse washers be hand-fed by others for several months. During one ceremony, Cook was told that certain baskets were tabu, but after the ceremony, they were not. Eventually, Cook understood that tabu also had a dual meaning, describing both the forbidden and the sacred.*

There's often a moral or religious reasoning behind taboos, whether in Tonga or anywhere else in the world. Some serve practical purposes as well, protecting against potential contamination, victimization, or other calamity. For example, eating right after touching a ripe corpse could make people sick; the Tongans noticed this, and made eating with one's hands taboo for months after contact with a corpse: perhaps inconvenient, but certainly effective.

Science can sometimes justify the logic of taboos, such as those involving corpses, by explaining their origins. But just as often, scientists find themselves in the position of challenging these cultural beliefs. Today, many decry genetic engineering as a path to creating "designer babies" and "Frankenfoods"; its practitioners are sometimes accused of playing God or meddling with nature (both entrenched Western taboos). Medical research using embryonic stem cells holds great promise for curing disease but comes into conflict with taboos around abortion. And in a nation where you can sign up to donate organs on your driver's license, it's easy to forget that organ transplants were controversial in their early days—and remain so in some parts of the globe.

....

* Indeed, the name of Tonga's main island, Tongatapu, is translated to English as "sacred south," not "forbidden south."

In Japan, for instance, the first legal* heart transplant wasn't performed until 1999, more than three decades after the world's first such operation, because of a taboo against taking an organ from a brain-dead donor whose heart was still beating: A person wasn't considered dead until the heart stopped.

In the last decade, we've seen a spate of other transplants that were formerly off-limits: hands, faces, and most recently, the world's first total penis and scrotum transplant, performed in 2018 on a soldier whose genitals were blown off by an improvised explosive device in Afghanistan.

For a long time, these kinds of body part swaps were verboten not only because of their riskiness, but also because they violated body taboos. Swapping a kidney or liver that remains unseen inside the body is one thing. But seeing one person's hands or face on a different person's body (much less having sex with another man's penis) was disturbing to many people. It just felt *wrong*.

Today, these kinds of highly visible transplants are becoming more common and accepted. (In 2018, *National Geographic*'s cover even featured a woman who received a face transplant.) But even as taboos evolve, they haven't all been swept aside. When doctors at Johns Hopkins performed the penis and scrotum transplant, there was one hitch: The scrotum was empty. It seems surgeons weren't willing to break one sex taboo: transplanting testicles. They consulted with bioethicists and determined that if they did, the recipient could possibly make babies with another man's sperm, babies whose biological father would technically be a dead man. And that, the Hopkins team decided, was a bridge too far.

The same is true of the head transplant, which (though it could theoretically save lives and extend life span) has never been attempted on a human being. A few years ago, I met a gravely ill

---

* As it happens, there had been one *illegal* heart transplant there, in 1968. The surgeon who performed it was investigated for the murder of the brain-dead donor.

Russian man, Valery Spiridonov, who volunteered to have his head relocated to a new body. Spiridonov suffers from Werdnig-Hoffmann disease, a progressive and eventually fatal muscle-wasting illness, but his mind is perfectly intact. The Italian doctor who proposed to do the surgery, Sergio Canavero, proposed putting Spiridonov's head onto the healthy body of a brain-dead donor. Canavero sees himself as a medical savior, bravely breaking taboos to battle death itself. Plenty of other people think his proposal is monstrous.

A head or brain transplant steps into a morass of ethical issues, beginning with the problem of obtaining consent from a brain-dead body donor. (When people sign up to donate their bodies to science, they probably aren't imagining that someone else's head might walk off with their body.) Then there's the question of who this new, combined person is. Does identity follow the brain, or the body—or both?

But Canavero is right about one thing: We had better start thinking about which taboos we're willing to break in the name of progress. Eventually, someone might attempt a head or brain transplant on a human—as in 2018, when a Chinese scientist named He Jiankui edited the genes of twin girls while they were embryos, making them resistant to HIV infection. The world's first gene-edited babies were born to international outcry. (Editing the DNA of human embryos is considered a first step toward creating genetically superior "designer babies.")

It's worth noting that important parts of the Crispr method that Jiankui employed in 2018 were first published in 2013; therefore, only five years had passed before the biggest taboo related to its use had been broken. The same thing could potentially happen with extreme transplants. After all, although a head transplant presents enormous technical and logistical challenges, it doesn't require a technological breakthrough equivalent to inventing gene editing. It could be attempted, as Canavero envisions, using existing surgical procedures and antirejection drugs.

Were that to happen, there would certainly be plenty of ethical, spiritual, and practical objections. But if history is a guide, it's hard for science to close a door once it's been opened. The announcement of the world's first gene-edited babies came as scientists gathered at the Second International Summit on Human Genome Editing; there, they condemned the editing of the girls' DNA, but also voted against a complete moratorium on human gene-editing research. After all, the technology has the potential to prevent deadly diseases such as Huntington's, an inherited disease that slowly kills off brain cells. Likewise, if a head transplant *did* extend someone's life, it might be hard to dissuade other gravely ill people from taking the risk. So now, we need to decide just how far we're willing to go to save a life.

And that's far from the only taboo that's starting to feel like a moving target. Whether regarding sex, death, or food, people around the world are grappling with which prohibitions serve a useful social function, and which are starting to look quaintly outdated. Taboos against murder aren't likely to be tossed out anytime soon, though as we'll see later in this chapter, science is providing new context for how much of our collective violent tendencies are "natural," given our evolutionary ancestry. Cannibalism and necrophilia are still powerful taboos as well—although, as with murder, we've come to understand that both of these are far more common across the animal kingdom than we had previously realized.

Meanwhile, other taboos—particularly those surrounding sex and women's bodies—are getting shakier. For millennia, pretty much anything having to do with lady bits has been considered off-limits to some degree. In parts of South Asia, women have long been banished from their kitchens or even their homes while menstruating, and Western medicine long gave women such short shrift that some female anatomical parts didn't even appear in textbooks. As we'll see, taboos about women's bodies are now changing, thanks in part to more women in science and medicine who are taking an interest in studying female physiology on its own terms. Little by little—or should I say lady bit by lady bit—we're chipping away at the taboos that keep women's bodies cloaked in mystery.

And science has so many more taboo lands left to explore. Whether looking at sex, cannibalism, or murder, scientists inevitably get curious enough to push our boundaries.

IT'S HARD TO GET A HEAD

Inside the world's most taboo operation

Valery Spiridonov looks impossibly small. He's dressed all in white, from his button-down shirt to his pants to the socks on his feet. A white blanket rests on his lap. Breaking up the look is a black strap, which holds him to a motorized wheelchair. Spiridonov's condition is grave: He has a rare degenerative motor neuron disorder called Werdnig-Hoffmann disease that is slowly killing him.

Spiridonov uses his left hand, which he can still move, to steer the wheelchair into a hotel meeting room. There, he confirms that he would like to be the first person to have his head transplanted onto a new body.

It's June 12, 2015, in Annapolis, Maryland, and Spiridonov, 30, has flown from Russia to attend the annual meeting of the American Academy of Neurological and Orthopaedic Surgeons (AANOS). He is there to support the surgeon, Sergio Canavero of Turin, Italy, who has proposed to do the transplant and is giving the conference's keynote address.

The meeting is small, maybe 100 or fewer surgeons, and held in an ordinary Westin Hotel. I arrived to find Canavero conducting a small impromptu press conference in the lobby, much like a coach amping up the crowd for a big game. Conference organizer Maggie Kearney has spent much of the day turning away reporters in anticipation of a packed room. She says that in 15 years, she can't remember a reporter attending this event.

During his talk, which lasted almost three hours, Canavero reviewed the scientific literature on spinal cord injury and recovery, as well as the potential for regrowth of various parts of the central nervous system. He asserted his belief that some of the basic assumptions of neurosurgery are wrong. Occasionally, he would point to Spiridonov, his wheelchair parked near the stage, and make a declaration ("Propriospinal tract neurons are the key that will make him walk again!"). By the end, most of the reporters in the room seemed worn out.

Responding to detractors' comments that the transplant could be "worse than death" or could drive his patient insane, Canavero asked Spiridonov directly, "Don't you agree that your [current] condition could drive you to madness?" Spiridonov answered quietly in the affirmative.

"I am sure that one day gene therapy and stem cells will fulfill their future," Canavero said, "but for this man it will come too late."

Finally, near the end of the talk, Canavero outlined the surgery at hand. He plans to sever the spinal cord cleanly, using a special scalpel honed nano-sharp. (I could not see Spiridonov's reaction to this detail, but wondered.) To minimize any die-off of cells at the severed ends during the transfer, Canavero will cut Spiridonov's spinal cord a bit lower, and the host body's a bit higher, then slice them again for a fresh cut. Next, he'll add some polyethylene glycol, which has been shown to stimulate nerve regrowth in animals, join the two ends together with a special connector—and voilà! Electrical stimulation would encourage further regrowth.

In this transplant scenario, repairing that severed spinal cord is an important key to allowing one person's brain to control another's body; signals must be able to pass from the brain to the body, and vice versa. Of course, there's a bit more to it, like reconnecting all the blood vessels, keeping the brain oxygenated during the process, and so forth. But Canavero is a neurosurgeon; the spinal cord is his focus. He points to the success of face and hand transplants, as well as other microsurgeries, in reconnecting nerves and muscles to restore movement.

The big medical question, it seems, is not so much whether the plumbing of two bodies can be hooked together, but whether the resulting person would survive for long or have anything resembling an acceptable quality of life. And, of course, there's the problem of tissue rejection, which would require a lifelong regimen of antirejection drugs that suppress the immune system.

Yet Canavero has no shortage of confidence in his efforts. He invited surgeons at the meeting to join his team, which could be enormous—more than 100, he has said—and plans to assign team leaders in orthopedics, vascular surgery, and so on. These surgeons would work on the project full-time for two years and "be paid through the nose, because I think doctors involved in this should be paid more than football players." (So far, though, it doesn't appear that these big bucks have materialized.)

For their part, neurosurgeons responded cautiously to the proposal. Such a surgery might be possible, "someday, but it is really a delicate situation," said Kazem Fathie, a former chair of the board of AANOS. Craig Clark, a general neurosurgeon in Greenwood, Mississippi, calls Canavero's idea provocative. "There have been many papers over the years that have shown regeneration, but for one reason or another they didn't pan out when applied clinically," he observed.

Although neurosurgeon Quirico Torres of Abilene, Texas, concedes that the scenario presents many ethical questions, he thinks it could make sense to allow volunteers to participate in the surgery; one day we might consider it normal. "Remember, years ago people were questioning Bill Gates: Why do you need a computer?" he asks. "And now we can't live without it."

After the talk, Spiridonov disappeared into a room to rest. When he came out, he faced the TV crews that had descended, sounding a bit weary of answering the same questions. "What will happen to you if you don't get this surgery?" a reporter called out. "My life will be pretty dark," he said. "My muscles are growing weaker. It's pretty scary."

While Spiridonov dealt with the press, I chatted with the hosts of his trip to Annapolis, who are his family friends. "He's brilliant, he's happy, he's funny," said Briana Alessi. "If this surgery were to go through and if it works, it's going to give him a life. It's life-changing. He'll be able to do the things he could only dream of."

And if not? "He's taking a chance either way," she said.

The final question posed to Spiridonov was this: What do you say to people who say this surgery should not be done?

His reply: "Maybe they should imagine themselves in my place."

As technological leaps go, head transplants would be a significant step toward one of Canavero's larger goals: radical human life extension. For decades, futurists like Ray Kurzweil have been mulling how to fight aging—and at least a couple hundred people, including baseball legend Ted Williams, have had their heads or bodies cryogenically frozen in the hopes that science will eventually vanquish death. Canavero favors keeping our brains alive in healthier bodies, and theorizes that one day we might even be able to clone brainless bodies and transplant our brains into them. If we cloned bodies genetically identical to our own, he suggests, rejection by a foreign immune system wouldn't be an issue.

But quite a few norms and taboos *would* be an issue. Any brain or head transplant raises questions about how we define death, and how we treat dead bodies. Canavero maintains that we need to get over the yuck factor for head transplants, just as we did for other parts, and that his surgery would be lifesaving for patients with terminal illnesses. As for the brainless clones, he wrote in the journal *AJOB Neuroscience,* "this is truly a matter for societal debate."

When I emailed with him a few years later, in late 2018, Canavero was as upbeat as ever about the prospects of carrying out his head transplant procedure—but it still hadn't happened. He explained that he's no longer sharing details about the project, nicknamed HEAVEN, with the media. He has moved his effort to China, which allows him to conduct research despite mounting global concerns about ethics of the entire proposition.

"Unfortunately," he wrote, "I am not free to share any details regarding HEAVEN for the reason that UNESCO, under pressure from the Vatican and several academic organizations, opposes our project . . . the publication of the first head swap in human cadavers published last year scared them all."

In another message, he mentions HEAVEN's next "evolution": brain transplantation—or more properly, a meningoencephalosomatic anastomosis. Anastomosis is the connection of two parts, often blood vessels—but here he's talking about moving the brain and its surrounding membranes (the "meningoencephalo" part) into a new body (the "somatic" part). Canavero offers no specifics as to how close he might be to attempting such a procedure, or whether he has a patient lined up.

As for his former patient, Valery Spiridonov, much has changed since that 2015 press conference, and much remains the same. For starters, his head is still on the same body.

When Spiridonov volunteered to have his head transplanted in 2015, he had already lived a decade longer than many people with his condition. And as 2019 began, he was alive and well. He remains wheelchair-bound, but in an Instagram photograph taken at the Hermitage in St. Petersburg, Russia, he looks stylish in a black-and-white leather jacket, one jeans-clad leg crossed over the other. In another photo, he gazes at a baby in a stroller, with the hashtags #familytime and #babyboy. That's right: Spiridonov is a father.

He bowed out of Canavero's plans for a head transplant in 2017, around the time Canavero moved his program to China. He still faces the possibility of a shortened life span due to his illness, which is genetic and incurable. But in April 2019 he told *Good Morning Britain* in a rare interview that his condition is stable for now, and that he has too much to live for to attempt a risky surgery. Along with a beautiful new wife, he has a healthy baby boy he calls a "miracle": The child did not inherit his disease. He and his growing family have moved to Florida, where he worked as a research assistant at Florida Atlantic University in software engineering. One of his projects was a wheelchair with autopilot. As I completed this book, Spiridonov was working as an engineer at a Florida technology firm.

In the end, Spiridonov's story raises as many questions about the head-transplant venture as it could have answered. If Canavero had gone ahead with the procedure in 2017, would Spiridonov have died on the operating table, or shortly thereafter? Or might he have survived, at least briefly, with hope of one day walking again? We'll never know. Is it better to find happiness, however fleeting, in one's own body than to seek a longer life in another? That's a decision that would be hard to make for yourself, much less for others.

If Canavero and Spiridonov can teach us one thing, it's that facing death is a unique and totally individual proposition. When it comes to our choices, we might throw taboos out the window, or we might look to tradition to guide us. The only universal is death itself; how and when to fight it is a question we all must face for ourselves.

A HEADY HISTORY

When Sergio Canavero announced his plans to transplant a human head, many people didn't realize that there had already been more than a century of quiet experimentation in the discipline. There had even been some measure of gruesome success— at least in the surgical sense.

1908: Charles Guthrie grafted a dog's head to another's neck, attaching arteries so that blood flowed first to the decapitated head, then to the original head. The transplanted head was without blood flow for about 20 minutes and regained only minimal movement.

1923: In Vienna, biologist Walter Finkler carried out a series of experiments transplanting the heads of insects onto each other's bodies, including cross-species head swaps. He asserted* that the new heads could control the bodies, and that male bodies given female heads began to show female behavior. Because insects' circulatory and nervous systems are much simpler than those of mammals, the "surgery" mostly involved cutting off the heads with sharp scissors and allowing the tiny drop of fluid that leaked out to cement the head to a freshly decapitated body.†

1950s: Vladimir Demikhov, a pioneer in human heart and lung transplants, grafted the upper bodies of young dogs onto the shoulders of other dogs, creating dogs with two heads, both able to move, see, and even lap up water. Without drugs to prevent rejection by the immune system, most lived only a few days (though one reportedly held out for 29 days). People weren't too happy about these experiments at the time. Even a tribute to the late Dr. Demikhov notes that it was unclear to many other surgeons what medical value the grafted dogs held.

1965: Robert White of Cleveland Metropolitan General Hospital transplanted the brains of six dogs into the necks of other dogs to show that the brains could be kept alive in another body.

---

* Finkler's experiments were discussed at some length in the journal *Nature* at the time. One experimenter, J. T. Cunningham of London, reported that he swapped the heads of mealworms and later found that the heads had died while the bodies lived on for several days. "The only remarkable thing in the result of these experiments was the tenacity of life of the insect body after decapitation," he wrote. However, some insect head transplants have worked; biologists today still do them to study the role of insects' brains in producing hormones and other chemicals that control metabolism and development.

† Kids, please don't try this at home.

The brains showed EEG activity and took up oxygen and glucose. (No telling what, if anything, they made of being trapped in an alien body.)

1970: Robert White transplanted the entire head of a rhesus monkey onto another monkey's body. According to White, the monkey could see, hear, and taste, but White did not attempt to fuse the spinal cord, so it was paralyzed from the neck down. The monkey lived for several days (reports vary from three to nine days).

2002: In Japan, scientists grafted the heads of baby rats onto the thighs of adults to test a method of cooling the brain to prevent brain damage with oxygen loss. The young rats' brains continued to develop for three weeks.

2013: Sergio Canavero proposed a human head transplant. In the journal *Surgical Neurology International*, he outlined a procedure involving a clean cut to the spinal cord to minimize damage and using polyethylene glycol, or PEG, to fuse the spinal cord.

2014: Canavero's collaborator, Xiaoping Ren, and his colleagues in China reported a head-swapping experiment in mice, resulting in a white mouse with a black head, and vice versa. The mice lived up to three hours after being removed from a ventilator. That's not long, but by keeping the donor body's brain stem, the body was able to continue to control its own heartbeat and breathing.

2015: Canavero detailed his head-transplant procedure. He proposed cooling the head and donor body to limit cell damage from oxygen loss, and fusing the spinal cord with a process he calls GEMINI, which uses PEG and electrical stimulation, shown in other studies to promote spinal cord repair.

2017: At the Harbin Medical University in China, a team led by Canavero and Ren transplanted the first human head— using cadavers. The surgery, which took 18 hours, was a practice run to work out the logistics of the procedure in living bodies. "This rehearsal confirmed the surgical feasibility of a human CSA [head transplant] and further validated the surgical plan," they reported.

THE MOST MURDEROUS MAMMALS

Our most prolific killers, revealed

Picture the most murderous mammal in the world. Not the best predator, taking down prey with a single swipe of a great talon or claw, but the one that excels in slaying its own kind.

Are you picturing a human being? Well, you would be wrong. But you might be surprised to know *Homo sapiens* actually falls at number 30 out of more than a thousand species on the list of animals that most often kill members of their own kind. Humans, it turns out, are just average members of a particularly violent lot, the primates. And the most prolific murderers* in the animal world are a different species altogether.

Which, you might ask? Believe it or not, it's the meerkat, a cute little African mammal belonging to the mongoose family and immortalized in the wisecracking character Timon in *The Lion King*. Sure, they look adorable when they stand up on their back legs to survey the savanna—but they're still vicious, baby-killing cannibals. About one in five meerkats (mostly infants) are killed by members of their own species, compared with just over one percent of humans whose deaths were linked to violence, whether murder or war. (For perspective, about 3 percent of the human population died during World War II, the deadliest conflict in history.)

Unlike humans, meerkats are not encumbered by murder or cannibalism taboos—or by notions of gender. In fact, most often

---

* Some grammar police get very exercised about using the word "murder" for nonhumans. I think they should lighten up. Here, it refers to one animal killing a member of its own species through violence—not predation, in which an animal kills a different species to eat it.

it's the females that kill. Within their matriarchal social system, meerkats maintain a strict hierarchy; the dominant female in a group usually prevents subordinate females from mating. If a subordinate gives birth, the alpha female will often kill—and eat—her pups, thereby reserving more resources for her own young. These raids are brutal, and I have to admit that despite my love for nature documentaries, I've watched only a few episodes of the popular *Meerkat Manor*. The violence stresses me out.

After meerkats, several varieties of primates top the murder rankings, including monkeys and lemurs. As many as 17 percent of deaths in some lemur species result from lethal violence. Some of the other top spots went to predators like lions and wolves—but plenty of species might also come as a surprise, such as sea lions, chinchillas, a species of gazelle, and squirrels.

The rankings are the result of the first in-depth survey of mammalian violence, led by José María Gómez of Spain's University of Granada in 2016. Gómez and his colleagues sorted through more than four million recorded animal deaths from 3,000-plus studies of 1,024 different mammal species and tallied the rate of lethal violence within each. Based on historical and archaeological records, the team also estimated the percentage of people who met violent ends throughout human history, as well as our potential for violence based on our evolutionary relationships (because closely related species tend to be similarly vicious).

One pattern stood out: Over the long run of mammal evolution, lethal violence increased. Although only about 0.3 percent of all mammals die in conflict with members of their own species, that rate is 2.3 percent in the common ancestor of primates. It would have been 1.8 percent in the first apes that evolved among the primates, Gómez's team calculated. Likewise, early humans would be expected to have had about a 2 percent rate of lethal violence; this lines up with evidence from Paleolithic human remains.

But if you zoom in to examine only humans—and in particular, the last few thousand years—violence doesn't just keep increasing. Instead, it has waxed and waned with the tides of human history. The medieval period was particularly brutal; if the plague, childbirth, or famine didn't get you, someone else might. Between bloody revolts, wars, and religious crusades, there were family blood feuds, brawls, and a general tendency to settle disputes with brutality. As a result, the era's records suggest that human violence was responsible for a whopping 12 percent of recorded deaths.

But since then, lethal violence has taken a downturn. For the last century, we've been relatively peaceable; according to a 2013 United Nations report, today's global homicide rate (which counts murders, not deaths in war) is just .0062 percent.

"Evolutionary history is not a total straitjacket on the human condition; humans have changed and will continue to change in surprising ways," says Gómez. "No matter how violent or pacific we were in the origin, we can modulate the level of interpersonal violence by changing our social environment. We can build a more pacific society if we wish."

Gómez's study also contains this feel-good nugget: Sixty percent of mammal species are not known to kill one another at all. Of more than 1,200 bat species, for example, very few kill one another. And apparently pangolins and porcupines get along fine without offing members of their own species.

Whales are also generally not known to kill their own kind. But biologist Richard Connor of the University of Massachusetts Dartmouth notes that in 2013, a pair of male dolphins were documented attempting to drown a baby right after its birth. Two or three male dolphins will often gang up on individual females, and it's possible that these males were trying to kill the infant so that its mother would become available for mating. He cautions that whales, as their close relations, might also be more violent than we've thought. "We could witness a lethal fight in dolphins but not know it, because the victim swims away apparently unimpaired, but is bleeding to death internally," he says.

More often, though, people think animals are more violent than they really are, says animal behavior expert Marc Bekoff, an emeritus professor at the University of Colorado, Boulder. He contends that humans and nonhumans alike are predominantly peaceful. But just as roots of violence can be found in our evolutionary history, he points out, so can roots of altruism and cooperation. Bekoff cites the work of the late anthropologist Robert Sussman, who found that even primates—some of the most aggressive mammals—spend less than one percent of their day fighting or otherwise competing.

That makes sense: Challenging another animal to a duel is risky, and for many, the benefits don't outweigh the risk of death. Highly social and territorial animals are the most likely to kill one another, which helps explain the trend in human history. Many primates fit that killer profile, though not all of them: Bonobos have mostly peaceable, female-dominated social structures, whereas chimps are much more violent.

These differences among primates matter, says Richard Wrangham, a biological anthropologist at Harvard known for his study of the evolution of human warfare. In chimpanzees and other primates that kill one another, infanticide is the most common form of killing. This often involves males killing unrelated infants to boost their chances of mating with the mother and passing on their own genes. But humans are different; they frequently kill each other as adults. "That 'adult-killing club' is very small," Wrangham says. "It includes a few social and territorial carnivores such as wolves, lions, and spotted hyenas."

Although humans may be expected to develop some level of lethal violence based on their family tree, that tendency is still surprising in other ways, Wrangham says. It's not just the rate of killing that matters, after all, but *why* animals kill.

In his 2019 book *The Goodness Paradox,* Wrangham points out two types of aggression, each with its own biology—and that either one can lead animals to kill. On average, humans are relatively low on the first type, known as reactive aggression. This is what Wrangham describes as the "hot" type, involving losing one's temper or crimes of passion; this is what we associate with fighting throughout the animal kingdom. It's a response to an immediate threat, and involves anger or fear. For the most part, people keep this kind of aggression in check. (If we didn't, road rage killings would be an everyday occurrence.)

But compared with other species, Wrangham argues, humans are high on the "cold" type of aggression, called proactive aggression. This kind has a goal—whether money, revenge, power, or something else—and it leads to calculated, deliberate murder. *Homo sapiens* are the masters of proactive aggression: It fuels terrorism, school shootings, even wars.

Proactive aggression can be found in nonhuman species, too, like chimpanzees conducting raids on rival troops. But as Bekoff observes, that's rare. "I hate when there's human violence and people say 'they're behaving like animals,'" he says. Most animals, including people, actually spend little time in conflict, he explains—but in nature specials, blood sells. His message: We can't use our lineage as an excuse for our own violence.

It's frightening to consider that humankind has used its fantastic intelligence not only to build complex civilizations, but also to perfect new and more efficient means of killing one another. But it has—and our exceptional capacity for proactive violence makes us unique, even if we aren't the most prodigious killers among mammals. And we can't just blame our "animal nature" for the violence we've cooked up. As Wrangham says, when it comes to murderous tendencies, "humans really are exceptional."

A PRACTICAL GUIDE TO CANNIBALISM

Why some animals put their own kind on the menu

Years ago, I worked at a nature center in Louisiana, where my job included caring for animals that were part of our public display and educational programs. This rotating menagerie mostly consisted of local fauna, such as baby alligators, opossums, and snakes.

Occasionally, other zoos and nature centers would loan us animals, including at one point a male and female pair of eastern screech owls. The tiny owls were adorable with their enormous eyes and fluffy bodies, and they could usually be found snuggled side by side. But one day not long after the owls arrived, I opened their cage at feeding time and discovered the male owl missing. I panicked, afraid that he had somehow escaped on my watch. I searched the room high and low—no owl. Then I noticed a lump on the female's chest.

Sure enough, she later regurgitated the indigestible bits of her mate as an owl pellet, and we ended up returning just one owl to the zoo. Cannibalism, it turns out, is not entirely unheard of in captive animals stressed by changes like a new environment.

The incident was unsettling in part because it seemed so incongruous. How could such a cute animal, seemingly so attached to her mate, gobble him down? But no law of nature trades beauty for bloodlust. The truth is that although most human societies consider it taboo, the animal world generally takes a more practical approach to killing and eating their own kind.

That applies even for animals we think of as harmless, like seahorses. Yes, the beloved seahorse—a denizen of Disney movies and little girls' sticker books—occasionally vacuums up its own young with its cute little snout. I could also ruin butter-flies and squirrels for you. And if you've ever raised bunny rab-bits—well, you may already know.

Animals turn to cannibalism for plenty of reasons, and some species make it a habit. Mothers who eat their young are pretty common; in fact, eggs and newborns are common victims of cannibalism in the animal kingdom. The reasons for this are complex; sometimes mothers eat their young because food is scarce, but research has shown that this isn't always the case. Some species overproduce eggs or babies and then winnow out the inferior specimens (such as those that don't develop quickly enough). But kids can become cannibals, too.

Some species of insects, scorpions, worms, and spiders prac-tice matriphagy, or mother-eating. One remarkable case occurs in crab spiders: Mothers provide their spiderlings with unfertil-ized "nurse" eggs to eat. The young eat the eggs—and also, slowly, their mother. Over the course of weeks, they gnaw at her until she falls immobile, at which point they consume her entirely. At least it's not for nothing: Mom-eating spider babies generally do quite well, with higher weights and survival rates than those of more respectful young. (Come to think of it, we're fortunate that more babies haven't hit on the idea.)

Luckily for us, evolutionary forces dictate that human mothers are safe. Only young that are born ready to care for themselves can pull off matriphagy—so even if human fetuses had teeth (and aren't you glad now that they don't?), a matricidal cannibal baby would chew its way out of Mommy only to find itself helpless and without a source of milk.

But even without cannibalistic kids, humans have certainly eaten each other from time to time—and we're not just talking about serial killers and starving plane-crash survivors. In fact, archaeologists have found evidence of cannibalism in the human family tree going back at least 800,000 years, in the form of prehistoric skeletons showing signs of being butchered and eaten in the same way as other animals. The question has been, Why? Do humans only eat one another out of necessity or as a symbolic ritual—as we might like to believe—or is it possible that we once hunted one another as food?

Some anthropologists have suggested that may be the case. But archaeologist James Cole of the University of Brighton wondered if that made sense. Does human flesh even have enough calories to make hunting one another worthwhile? When he looked for numbers, it turned out that for all our calorie counting, we had never turned the tables on ourselves to figure out how humans might rate as food.

So Cole did the calculations using detailed data on the composition of four adult males. He found a 145-pound human holds about 32,000 calories in skeletal muscle, compared with about 3.6 million in a mammoth or 163,000 in a red deer. He also came up with a rather macabre nutritional table listing the number of calories in various parts of the human body. (In case you're ever in the unfortunate position of needing to know, you might want to start with the thighs: Human thigh meat has more than 13,000 calories, compared with about 2,500 in the liver.)

It turns out, compared with other mammalian menu items, humans don't make particularly good prey—perhaps another reason why cannibalism isn't more widespread. Cole's research found that although a dead mammoth's muscles could feed 25 hungry Neanderthals mammoth steaks for a month, cannibalizing a human would provide the same crowd with only a third of a day's calories. (Another way to parse that figure is to understand that to a roving pack of 25 Neanderthals, you'd be lunch.)

And that's not just because of our comparatively meager size. Pound for pound, compared with other animals, we're not very nutritional. According to Cole's estimates, boars and beavers pack about 1,800 calories into each pound of muscle, compared with a measly 650 calories for a modern human. That's about half the calories in a pound of 80 percent lean ground beef.

So, if humans aren't especially valuable in terms of prey, why eat them? After all, unless they're sick or dying, even an ancient human wouldn't be easy to hunt. "You have to get together a hunting party and track these people; they aren't just standing there waiting for you to stab them with a spear," says Cole. Instead, he argues that perhaps not all ancient cannibalism was for the purpose of filling bellies. It may have also served various social functions for early humans and their ancestors.

In some cases, cannibalism may have been purely practical. "The issue is not one of nutrition as an alternative to large game," says anthropologist Erik Trinkaus of Washington University in St. Louis. "It's an issue of survival when there are no other food sources, members of one's social group have died, and the surviving members consume the bodies of already-dead people."

But in some situations, other scientists suggest, cannibalism might have been a choice. Though bones can't reveal motivations, ancient remains do offer a few clues to cannibalistic practices. At the Gran Dolina cave site in Spain, for instance, the butchered bodies of bison, sheep, and deer were mixed with those of at least 11 humans—all children or adolescents—whose bones showed signs of cannibalism. In addition to marks showing that flesh had been stripped from the bone, evidence suggests the Gran Dolina residents—an ancient human relative called *Homo antecessor*—ate their victims' brains.

In the cave, the butchered human parts appear in layers spanning about 100,000 years, suggesting that humans were a recurring menu item. The remains were also mixed in with those of other animals that had been prepared the same way, leading some anthropologists to suggest that cannibalism at the site might not have been done in a food-stress emergency or as ritual behavior. Perhaps human flesh was a common supplement to their diet. Or perhaps the youngsters were outsiders, and cannibalism served as an effective "keep out" sign? Sadly, the bones can't say.

That's true in most cases of prehistoric cannibalism, says anthropologist Silvia Bello of the Natural History Museum in

London. "Paleolithic cannibalism was probably more often practiced as a 'choice' rather than mere 'necessity,'" she says. "I think, however, that to find the motivation . . . is a very difficult matter."

Cole acknowledges we can take only so much from his limited analysis of human nutritional value, which was based on only a few modern specimens. And clearly, our ancient ancestors weren't counting calories or totaling the Weight Watchers points for bone-in human thighs when they made their dinner selections.

Perhaps, he says, the real message is that ancient people had a mix of motivations for cannibalism. After all, the practice in recent centuries has many roots in addition to mere survival, including warfare, psychosis, and spiritual beliefs.

At least we're not alone, as the screech owls I took care of remind us. "Cannibalism is extremely widespread in the animal kingdom," says Bill Schutt, a biology professor at Long Island University's Post campus and author of *Cannibalism: A Perfectly Natural History*. Most likely, he suggests, ancient people survived by being incredibly opportunistic, and occasionally cannibalistic, just like other animals. "What makes us different are the rituals, the culture, the taboos," Schutt says. "We've been patterned to believe that cannibalism is the worst thing you could do."

Thus, in most modern societies the practice is either forbidden or acceptable only in an emergency. It is sometimes done for reasons unrelated to hunger, such as the Amazonian Wari' tribe's historical practice of marking a relative's passage into death by eating a small piece of his or her flesh. Even then, it's not considered an appealing menu option.

For his part, Cole says that working out the calorie count of human flesh was a bit disconcerting. Apparently cannibals have described it as tasting like pork, which weighed on his mind as he did his analysis. "I found it quite difficult to eat bacon for the last year or so," he admits.

PIONEERING THE CLITORIS

*Exploring the body part that
didn't make the anatomy books*

Sixty-seven years after Christopher Columbus "discovered" the New World, another Italian named Columbus claimed the discovery of a territory far smaller, but no less wondrous. In 1559, Realdus Columbus proclaimed that he had discovered the clitoris.

In his seminal work *De re anatomica,* Realdus Columbus (or in Italian, Realdo Colombo) described a "certain small part" located above the opening of the urethra that was the main source of women's enjoyment of sex, and delighted in describing the effects of stimulating the spot.

Colombo was smitten with his discovery, and I imagine him inflicting his newfound knowledge on women with gusto. In any case, he was happy to lay claim to the organ, but knowing of no name for it, he suggested calling it the "love of Venus," or *amor Veneris.*

Just as Christopher Columbus's "discovery" was actually a continent* that Native Americans had inhabited for thousands of years, Colombo's supposed breakthrough now warrants a collective eye roll from the native inhabitants of the female body. But at the time, this was hot stuff.

Colombo wasn't the only 16th-century anatomist to stake his claim on the clitoris. His rival Gabriel Fallopius (better known

---

* It also wasn't the continent he's usually credited with discovering—Columbus never stepped foot in North America. He did lead expeditions to the Caribbean, Central America, and South America.

for discovering the fallopian tubes) was working on a book in which he snippily wrote that he had found it first, implying that everyone else (principally Colombo) was trying to steal the discovery. "If others have spoken of it," Fallopius wrote, "know that they have taken it from me or my students." Colombo had died by the time Fallopius published his book in 1561, but the battle nevertheless created a stir among European medical experts, with colleagues taking sides on the matter.

Of course, neither man was correct; at best, they had *re*discovered the clitoris. Not only had many satisfied women discovered it for themselves since, well, forever—but neither Colombo nor Fallopius was even the first to publish an anatomical study of this particular body part. The French anatomist Charles Estienne had described it in his own book in 1545, though he had scornfully referred to it as a "shameful member" and incorrectly ascribed it a urinary function.

Long before that, Greek, Persian, and Arabic physicians knew about the clitoris, though not by that name. Hippocrates called it the *albatra* or *virga,* meaning "rod." The Arabic writer al Zahrawi called it the *tentigo,* or "tension." The word "clitoris" didn't come into common use until the 17th century, and derived from an ancient Greek word, *kleitoris,* that's often translated as "little hill," or according to other sources, "to touch or tickle." (Or maybe it's both, and some ancient Greek wiseacre was making a pun.)

Whatever it was called, not everyone in Colombo's day was buying into the newfangled sex organ. Colombo himself had trained under the famed anatomist Andreas Vesalius, who dismissed the clitoris as an aberration found only in hermaphrodites. "You can hardly ascribe this new and useless part, as if it were an organ, to healthy women," he wrote. "I have never once seen in any healthy woman a penis . . . or even the rudiments of a tiny phallus."

And therein lies one reason the clitoris was so misunderstood. Vesalius held onto the belief, prevalent for centuries, that the female body was derived from the male's—the implication being that a woman's reproductive tract was simply the inside-out version of a man's. The vaginal canal was the inverted penis, the ovaries were the testicles, and the uterus was the scrotum. In Vesalius's view, the pieces were all accounted for without the troublesome clitoris.

What's more, the idea of women having their own source of sexual pleasure—much less one that did not depend on penetration—made some men deeply uncomfortable. In 1573, the French surgeon Ambroise Paré wrote of a "monstrous thing that occurs in the labia of some women" that could become erect "like the male penis, so that they can be used to play with other women."

Paré then related a story about how some African men put a stop to such female proclivities: They would go around "like our [livestock] castrators and make a career of cutting off those excrescences." He was talking about clitoridectomy, also known as "female circumcision," a form of female genital mutilation still practiced today in more than 30 countries.

Although clitoridectomies are most common today in a broad region across Africa and the Middle East, many people don't realize that the surgery was once more widespread, and was also employed in Victorian England. A successful surgeon named Isaac Baker Brown, credited with developing many gynecological surgeries, touted clitoridectomies as a cure for masturbation.* In his book *On the Curability of Certain Forms of Insanity, Epilepsy, Catalepsy, and Hysteria in Females,* published in 1866, Brown offered among his instructions that "the clitoris is freely excised by scissors or knife—I always prefer scissors." Eventually, the Obstetrical Society of London essentially put Brown and his clitoris removals on trial, ultimately

---

* Brown believed that masturbation caused a nervous disease that began with hysteria (a good old-fashioned catchall for misbehaving women) and eventually led to epilepsy and "idiocy." It's unknown how many women he subjected to this procedure.

expelling him from the society. But the practice lingered for decades, and spread to America. The last known clitoridectomy in the United States was performed on a five-year-old girl in the 1940s.

It's easy to see why, mired in debate over its very existence and treated as taboo when it was acknowledged at all, the clitoris languished for so long in the shadows of medical study. It would be another two centuries before a German anatomist, Georg Kobelt, would make major advances in clitoral description based on detailed dissections in the 1800s. But even well into the 20th century, some anatomy textbooks left the clitoris out of their diagrams entirely. Others showed only the external portion, ignoring the clitoris's deeper structure.

Perhaps its exclusion from textbooks is one reason why most people have never learned what the clitoris actually looks like, or just how big it is. Maybe you, as I once did, think of it as a pencil eraser–size organ; my friends, that's just the proverbial tip of the iceberg.

The sensitive bit that we can see outside the body is the glans of the clitoris, a nerve-rich bit of tissue analogous to the glans of the penis. Like the penis, the glans of the clitoris is attached to a shaft, which in the female case takes a right turn from the glans and disappears into the body. That's where many people would assume the story ends—or that at most, there's some short stalk beneath the skin. But no. Behind the glans, the shaft or body of the clitoris is about an inch or inch and a half (two to four centimeters) long, and then it branches out to form two arms, called the crura, that give the clitoris the shape of a turkey's wishbone. Those arms can be up to three and a half inches, or nine centimeters, long. And we're not done yet. Inside the wishbone are two elongated bulbs of erectile tissue that wrap around the urethra and vagina as if giving them a hug. The pea-size glans is not even 10 percent of the whole structure.

With its long, curved arms and pendulous bulbs, the clitoris reminds me of a flower's sex parts, or maybe half an octopus. And, if I'm completely honest, it also reminds me of the alien spaceships in the 1953 *War of the Worlds* movie, which had a boomerang-shaped body topped by a curved neck and a glowing, hooded head concealing a deadly ray gun. I suppose I should go with the flower analogy, though I also rather like the idea of a clitoris with a ray gun. In any case, it was nothing like what I'd imagined.

It was embarrassing to have reached the age of [cough, mumble] without knowing that there was so much more to the clitoris than met the eye. But the more I thought about it, it made me angry. Why didn't anyone tell me any of this? Well, most likely, none of my biology teachers knew more than I did. The anatomy I've just described was not published in some now yellowing treatise during the age of the great explorers. No, the first scientific paper to fully describe the anatomy of the clitoris—including the bulb structure, its nerves, and blood vessels described in detail—was published in 1998.

The woman responsible for this breakthrough in clitoral anatomy is Helen O'Connell, who was also the first female urologist in Australia. She's a busy surgeon in the urology department at Royal Melbourne Hospital, repairing the delicate tissues of the lower urinary tract so that men and women can pee, or not pee, whichever is the problem (her specialties are incontinence and urinary obstructions).

During the course of her training, O'Connell noticed that surgeons would carefully avoid the nerves and blood vessels important to men's sexual function during prostate surgery. But the same wasn't true during pelvic surgery on women. When O'Connell searched textbooks for details on the nerves and vasculature of the clitoris, they simply weren't there; there was no clear map of a woman's sexual anatomy, and therefore no way to be sure she wouldn't damage it.

O'Connell set out to fix the problem. She performed her own dissections of cadavers and compared what she saw with historical anatomical texts. Much of what Kobelt had drawn in the 1800s held up well; he just hadn't connected all the parts into one complete structure. Worse, the details he described were dropped from anatomical diagrams over the years. (In its 1948 edition, for example, the classic textbook *Gray's Anatomy* omitted the clitoris.)

So in 1998, O'Connell published "Anatomical Relationship Between Urethra and Clitoris" in the *Journal of Urology,* showing that the clitoris was larger and more complex than what most textbooks depicted. Not only was the bundle of nerves far larger than what *Gray's Anatomy* described, but when O'Connell dissected cadavers, she found that the two large pieces of erectile tissue that hug the vagina (mysteriously labeled in textbooks as the "bulbs of the vestibule") were directly attached to the clitoris, so she renamed them the bulbs of the clitoris.

I spoke to O'Connell about her work via Skype from her office in Melbourne. She's friendly and practical—just the kind of person who could make a patient feel comfortable as she peered at his or her urethra. She pushes a stray lock of blonde hair behind her cat's-eye glasses as we discuss the pronunciation of the word "clitoris." She puts the emphasis on the first syllable, not the second. "That's funny, isn't it," she says, "because there was an episode of *Seinfeld* where he talks about something that rhymes with Dolores. I suppose that's the English version versus the American version." Tomato, tomahto.

As O'Connell continues to pursue her clitoris research along with her surgical practice, her colleagues have taken notice. In 2005, she published an even more comprehensive review that included modern magnetic resonance imaging (MRI) studies and descriptions of the microscopic structure of clitoral tissue. The research hinted at a possible answer to an age-old sexual mystery: the source of the so-called vaginal orgasm. Many people believe that the completely vaginal orgasm, with no clitoral involvement, is a myth, a unicorn, a beautiful but impossible thing. The vagina just isn't sensitive enough.

The debate is wrapped in layers of sexism and not a small dollop of anger—much of it directed at famed psychoanalyst Sigmund Freud. In 1905, Freud asserted that healthy women move from clitoral-triggered orgasms to vaginal orgasms as they mature, and that clitoral orgasms are thus an infantile, immature phenomenon. If a woman couldn't reach orgasm through vaginal penetration alone, there was something wrong with her.

Eventually women got fed up with men telling them, in the guise of science, how to climax. Yet some women continued to report that they do *have* vaginal orgasms, which begs the question: What are they experiencing, and how? Enter the G-spot,* an area on the anterior part of the vaginal wall (meaning toward the front of the body) that some women report to be sensitive enough to induce an orgasm during intercourse.

O'Connell launched a search for the G-spot, dissecting 13 cadavers. In 2017, she reported that she couldn't find a sign of such a structure in the vaginal wall. But in her 2005 study, she had described how the erectile tissue of the clitoris wraps around the urethra and vagina, and the place where the three meet is indeed on the anterior wall of the vagina. It's possible, she wrote, that what we've been calling the G-spot is actually this complex where the three meet, and is stimulated by pressure through the vaginal wall.

To test the idea, someone needed to see what actually happens during sex. So in 2010, a research team in France recruited two medical doctors who were already a couple, and used ultrasound—the same equipment used for a sonogram of a fetus—to monitor the woman's vaginal wall and clitoris during intercourse.

It probably wasn't the best sex of the couple's life, because it happened behind a curtain in a gynecology office, but it offered the best look scientists could get at the clitoris during

....

* The G-spot was named for Ernst Gräfenberg, a German researcher who first reported its existence in the 1940s.

intercourse. And voilà: They reported that, "during intercourse, the anterior vaginal wall was crushed against the root of the clitoris." It appears the source of the vaginal orgasm may indeed be the clitoris—just the other end of it. The exact size and shape of the organ can vary among women—plus these structures move around during sex, which would help explain why the mysterious G-spot doesn't seem to work for everyone.

Depending on how you look at it, the G-spot either never existed in the first place (because it's the clitoris) or has been vindicated (because it's where we thought it was). Some defenders still believe it's a physical structure in the vaginal wall, and naysayers don't believe it exists at all. Yet in popular culture, you could say O'Connell's idea is starting to penetrate. Magazines like *Cosmopolitan* have started instructing readers on the relationship between the G-spot and the clitoris—but for most people, the proof will be in the pleasure (or lack thereof).

Meanwhile, O'Connell's anatomy studies have spurred their own sexual revolution, and others use her work to educate men and women about sex. An education on female anatomy is certainly in order. In a study of midwestern college students in 2010, more than a quarter of women and nearly half of men could not identify the clitoris on a diagram, even when given a list of words to match to the body parts.

So in 2012, artist Sophia Wallace started the wonderfully named Cliteracy project, a series of artworks that show off the clitoris in all its glory. And even textbooks have started to get on board. In 2017, French education activists pushed the nation's biggest publishers to include an anatomically accurate clitoris in a diagram of female anatomy. The change came after a six-year public awareness campaign that included mowing a nearly 400-foot-long outline of an anatomically correct clitoris into a farmer's field outside Montpellier, in the South of France. It looked just as O'Connell had described it, even if the local press was a bit surprised to find that the shape was a clitoris and not, as they described it, half a garlic clove.

Little by little, the clitoris is starting to get the recognition it's due—and it's only taken medicine a couple thousand years to figure it out. But each time it's "discovered," we get closer to actually understanding this crucial organ. Maybe one day, if all goes well, we'll recognize its shape as immediately as we do that other sex symbol: the phallic one.

ONE GIANT LEAP FOR WOMANKIND

Why menstruation is having a cultural moment

ally Ride's tampons might be the most discussed menstrual products in the world. Before Ride became the first American woman in space in 1983, scientists pondered the tampons she would take to the final frontier. They weighed them, and NASA's professional sniffer smelled them—better to take deodorized or nondeodorized?—to make sure the odor would not be overpowering in a confined space capsule. Engineers considered exactly how many she might need for a week in space ("Is 100 the right number?" they famously asked her).

The engineers were trying to be thoughtful; reportedly they packed the tampons with strings connected so that they wouldn't float away. I imagine Sally Ride's tampons hovering like sausage links in the space shuttle, and wonder if any unwitting male astronauts ever came across them and floated away, embarrassed.

Despite best efforts, it's safe to assume menstruation made the men of NASA squirm. When women became astronauts, there were not only the typical concerns that they would become weepy or unable to function during their periods—but also that the menstrual cycle might somehow break in space. Would the blood come out without gravity to pull it from the womb? Maybe it would all pool inside, or even flow backward through the fallopian tubes into the abdomen—a frightening condition called retrograde menstruation.

In the end, someone* just had to try it and see. And what

* NASA has never said whether that someone was Ride.

happened was . . . nothing much. The uterus is pretty good at expelling its lining sans gravity, as it turns out (after all, lying down doesn't seem to matter much). Dealing with space tampons is probably something of a nuisance, and space cramps can't be any nicer than Earth cramps. But NASA ultimately decided that menstruating astronauts were perfectly safe to fly.

Fast-forward to today, three decades after Sally Ride's tampons made the trip into space. Scientists have raised a possibility for female astronauts that has occurred to other women as well: Maybe we don't need to have periods at all. We have the technology. A combined oral contraceptive, or what we commonly know as the Pill, can be used continuously, without taking a week off to induce bleeding. That's currently the best and safest choice for astronauts who prefer not to menstruate during missions, says Varsha Jain, a gynecologist and visiting professor at King's College London. In 2016, she and her colleague Virginia Wotring, who as NASA's chief pharmacologist was asked to suggest the best contraceptive[*] for astronauts, published a study of space menses in the journal *NPJ Microgravity*. Contraceptive implants and IUDs are options, too, but the Pill already has a good track record in space.

As it happens, not only have female astronauts tried the continuous-pill method (to much less fanfare than Sally Ride's space tampons), but more women on Earth are opting out of periods as well. Polls suggest that about a third of women feel they need to have a monthly bleed because it seems natural and reassures them they're not pregnant. But the bleeding that occurs during the week off the Pill isn't necessary or even particularly natural, Jain says. Women on the Pill don't build up a uterine lining that needs to be shed. And having a flow doesn't ensure you're not pregnant.

....

[*] Not because NASA is necessarily concerned about safe sex in space; they just don't want women getting pregnant right before a mission. As for sex during a mission, there have been rumors, especially after the first married couple went to space together in 1991. But officially, NASA says it has no knowledge of anyone having sex in space, and has not conducted any experiments involving sexual activity.

Of course, the Pill itself does come with some risks: Blood clots in the legs and lungs are concerns, and its effect on cancer risk is mixed. According to a large-scale study published in 2018, the risk of breast and cervical cancer is possibly higher for women on the Pill, but risk for ovarian and endometrial cancers is lower. And although research hasn't found increased health risks from taking the pill for an extra week each month, there's some evidence that having periods could be good for the uterine health of women who plan to get pregnant. In the end, it all depends on your own set of pros and cons.

For long-term space travel, there are added benefits of skipping the flow. "The waste disposal systems onboard the U.S. side of the International Space Station that reclaim water from urine were not designed to handle menstrual blood," Jain and Wotring write. A woman spending three years in space—say to go to Mars and back—would need about 1,100 pills, which adds a little weight to a mission, but is far less unwieldy than all those tampons.

It's reassuring to see that NASA is getting a bit more comfortable with women's bodies when it comes to space travel. After all, if we ever colonize another world, we'll need to think a lot more about the mechanics of space reproduction. And even if we don't, more women will be heading into space, just as they've merged into other workforces. Since Sally Ride's "gynaugural"* U.S. mission, more than 40 American women have gone to space. And though historically only 11 percent of all space travelers have been female, women are slowly starting to catch up: In 2013, the astronaut class was 50 percent female for the first time.

Although women in increasing numbers are going to space, fighting in wars, and competing in Olympic sports—all while

---

* Sally Ride was the first American woman in space, but not the first woman in space. In 1963, Soviet cosmonaut Valentina Tereshkova took that title; nearly 20 years later, in 1982, Svetlana Savitskaya became the second woman in space. Sally Ride was third.

menstruating—periods nevertheless remain a taboo subject in varying degrees around the world. In the United States, mentioning your period in public is still usually treated as TMI; tampon advertisers make only veiled allusions to what their products are used for. In parts of India, it's taboo for women to prepare food during their periods because they're considered unclean. And only in 2018 did Nepal outlaw its *chhaupadi* practice, which banishes menstruating women to huts (often just a dirty lean-to). There, they are forbidden from touching other people, but are themselves vulnerable to rape or even murder. Advocates for women in Nepal fear the ban will be ignored.

There's a new generation, though, that's pushing back on menstrual taboos. Not only are people talking about periods more, but there's even a term now for a lack of menstrual education and hygiene: period poverty. And people around the world are working to end it. One of them is India's Pad Man. Arunachalam Muruganantham saw his new wife hiding something behind her back and asked to see it. "It's none of your business," she said, but he darted behind her. She was holding a dirty rag that she was using as a menstrual pad, because store-bought pads were too expensive.

Muruganantham decided he could make her something better. After dissecting pads to figure out their composition, he invented a low-cost machine that makes simple sanitary napkins for less than a third the cost of commercial pads. When he couldn't get any women to try them out, he tested them himself. Muruganantham put animal blood in a bottle he could wear at his hip, he explains in a TED Talk. "There is a tube going into my panties, while I'm walking, while I'm cycling," he says. "That five days I'll never forget. The messy days, the lousy days, that wetness." The TED audience roars with laughter.

Today, he's making his technology available to rural women so they can make affordable pads; hundreds are in operation. In 2014, Muruganantham was one of *TIME* magazine's 100 most influential people of the year, and the Bollywood film *Pad Man* chronicles his adventures.

"My vision is to make India into a 100 percent sanitary-pad using country," Muruganantham told the BBC. "Menstruation is no more a taboo." I'm not sure he's entirely right about that, but I hope he will be soon.

NE-CROW-PHILIA

Making sense of the most sordid sex in the animal kingdom

It was a beautiful spring day in Seattle, the cherry trees tufted in pink, as two scientists and a documentary film crew assembled to watch a dead bird's funeral. The bird, in this case, was a crow, and the scientists told the film crew what to expect. Normally, crows swarm around a fallen compatriot and caw loudly, a funerary behavior called mobbing that the scientists had observed often. But that's not what happened this time.

Instead, the first crow on the scene lifted its tail, dropped its wings, and strutted toward the feathered corpse. One of the scientists watching was Kaeli Swift, a graduate student at the University of Washington. As Swift later wrote on her blog, she was confused; the bird's swagger looked like mating behavior, not the usual alarm call. Then the unthinkable happened: The live crow mounted the dead one and began frantically copulating with it.

"Is it CPR?" a crew member asked. No, the scientists explained. They had just witnessed crow necrophilia,* or sex with the dead.

Necrophilia is perhaps the most horrifying of taboo acts. Not only does it violate the dead, but our taboo against it also seems to have a clear biological and evolutionary basis: Sex with a dead body risks disease and obviously isn't useful for reproduction. And yet much of the animal kingdom isn't squeamish about it; a long list of animals

....

* Some will argue that necrophilia is an attraction to dead bodies (with the "-philia" in "necrophilia"), and that an animal's copulation with a dead body is not born of such a psychological state and thus is not necrophilia. However, biologists also use the word "necrophilia" to refer to the actual act of sex with a dead body—and setting animal emotional states and predilections aside, I'm using the term in that sense.

have been caught in the act, including a variety of birds (penguins[*] and quite famously, mallard ducks,[†] among others), as well as frogs, lizards, primates, and at least one sea otter. And of course, the occasional *Homo sapiens*. It's assumed to be rare, because it's usually[‡] a waste of reproductive energies. But in reality no one knows exactly how common necrophilia is in the animal world—with the exception of crows, thanks to Swift's experiments.

Swift didn't set out to become an expert on crow necrophilia. But that spring day in 2015 changed everything. She was well into her Ph.D. research on crows' behaviors around death, and had already demonstrated that the birds' noisy funeral gatherings serve at least in part to alert the group of potential dangers. From what she had observed, there was generally some obvious danger nearby when crows formed a mob; live crows would maintain a safe distance from the dead.

But this necrophilic crow had thrown a wrench into Swift's research. Why did it try to mate with its deceased comrade instead of calling its buddies? Did it not recognize the other bird as dead? Was it somehow confused by the absence of a predator?

"My adviser had been studying crows for 30 years, and he had never seen necrophilia before," Swift told me.

It's difficult, to say the least, to surmise what an animal is think-

---

[*] Despite their adorable looks, penguins are some of the most sexually libertine animals on Earth. Males, facing a short mating window in a harsh environment, are extremely driven. In 1911, Antarctic explorer George Murray Levick described male penguins engaging in necrophilia, rape, sodomy, and masturbation. Levick's Edwardian sensibilities were deeply offended. "There seems to be no crime too low for these penguins," he wrote. His report was considered so scandalous that London's Natural History Museum wrote "Not for Publication" on it and shelved it for a century.

[†] In 1995, Dutch biologist Kees Moeliker observed a male mallard copulating with another mallard that had just died after flying into a window. The event went on for 75 minutes, and resulted in a paper on the first recorded case of homosexual necrophilia in mallards. The report garnered Moeliker an Ig Nobel Prize ("for achievements that make people laugh, and then think"). Later, it made for quite an entertaining TED Talk.

[‡] But not always. In one frog, an Amazonian species named *Rhinella proboscidea*, scientists have observed "functional necrophilia." A male mounts a dead female and squeezes her body until she releases eggs, which he then fertilizes. Because frogs normally release their eggs into the water anyway, the fertilized eggs can grow into tadpoles, making the sex a "functional" act of reproduction.

ing. Swift and her graduate adviser, John Marzluff, designed careful (if strange-looking) experiments to determine whether the mobbing behavior might confer some advantage to crows. In the experiments, volunteers approached groups of crows holding a taxidermied counterpart to see how they would react. Their reaction was usually to mob the volunteers, loudly scolding and sometimes dive-bombing the cadaver-carrying humans.

Making the scene even more surreal, each volunteer wore the same latex mask when approaching the crows, which not only looked creepy, but also made it appear that the same person kept showing up with dead crows. Weeks later, the crows would mob and scold anyone wearing the mask, even if he or she wasn't carrying a taxidermied bird, suggesting they had learned that face as a threat. After trying variations involving dead pigeons and live-looking taxidermied hawks, Swift and Marzluff concluded that crows do take special note when members of their own species die as opposed to others—and as suspected, they use the "funeral" behavior to warn the group about nearby predators, like the hawk.

Knowing all this, Swift wanted to figure out why, and how often, crows have sex with their dead. This required an even more bizarre set of experiments. This time, Swift enlisted a fellow student to learn some taxidermy skills and to prepare crows in both "alive" and "dead" poses. Then, each spring and summer for three years, Swift presented hundreds of crows with her taxidermied creations.

Unlike in previous studies, no "predator" was lurking near the dead birds this time; more crows were emboldened to approach and touch the cadaver. Twenty-four percent of the crows pecked or pulled at the corpse, sometimes even dismembering it. They were so violent, in fact, that Swift went through dozens of taxidermied specimens and dead birds obtained from a local wildlife rehabilitation facility.

It turned out that necrophilia occurred in about 4 percent of Swift's trials. The incidents occurred almost entirely during the early part of the birds' breeding season in spring, and Swift thinks the behavior likely has to do with higher hormone levels. "That's when they're very territorial and have the most to lose," she says.

Essentially, a crow is alarmed about finding a dead crow, but at the same time, its hormones are making it more aggressive and sexually aroused. It seems that for a few birds, the wires get crossed.

As for the seemingly simpler explanation that this is all merely a case of mistaking a dead bird for a live one, Swift doesn't think so. The crows in her experiments treated their taxidermied counterparts differently, depending whether they were posed upright (as though alive) or sprawled out flat (as though dead). They went through their funeral routine only if the bird appeared dead, suggesting that they know the difference and probably didn't try to mate with a dead bird out of ignorance.

When it comes to other animals, who knows? Crows are unusually smart and observant. Maybe other animals don't have the faculties to recognize an animal as dead, Swift says. It may depend on the species. Smart, social animals like elephants, dolphins, primates, and corvids (crows, ravens, and their relatives) stand out as animals that take special note of death and even appear to mourn the loss of close fellow animals. Their grief isn't always pretty—chimpanzee mothers have been known to carry around their dead babies for days or weeks—but it's clear they're feeling something intense.

For our part, we humans tend to get judgmental when other animals don't adhere to our prescribed lifestyles. We celebrate animals like the albatross, which form lifelong bonds with their mates, and overlook the fact that they sometimes "cheat" on their partners. And we definitely don't like to think of tuxedoed penguins shagging corpses.

But this says more about our discomfort with our own moral turpitude than it does about any kind of natural order. We would like to believe that our values are somehow stitched into the universe, like a homespun sentiment embroidered on a pillow. But the universe is not so kind—instead, it's up to us to set our rules and abide by them.

Of course, people sometimes break those rules. In the case of necrophilia, our motivation for doing so is even more inscrutable than that of crows. The most famous cases, though not the most common, involve murderers who defile their victims' corpses; that macabre list includes the notorious serial killers Jeffrey Dahmer, Ted Bundy, Ed Gein, and Ed Kemper. But I think one of the strangest cases, and perhaps even more difficult to understand, is that of radiologist Carl von Cosel (who also went by Carl Tanzler).

In 1926, von Cosel left his wife and daughter behind in Germany to sail for Key West, where he fell in love with a dead woman. Well, she wasn't dead when he met her. Maria Elena Milagro de Hoyos (or Elena, as she was known) was a 22-year-old knockout with mesmerizing eyes—and tuberculosis. She died of the disease in 1931, and von Cosel was so distraught that he paid for a tomb so that he could conveniently visit her. He sat in the tomb every day, and supposedly installed a phone line inside to "talk" to her.

But even that wasn't enough. Needing to stay in physical contact with Elena, he stole her body from the tomb under cover of night and took it home. As her embalmed remains deteriorated, he patched up her skin with silk soaked in wax and plaster and laid her out on his bed, fully dressed and bedecked in jewelry, with a wig made from her own hair.

Von Cosel kept the body for seven years before it was discovered and taken away from him. By then, the statute of limitations had expired on his grave robbing, and he went free. Twelve years later, he died alone at home, a life-size Elena doll by his side. Years later, doctors alleged that the real Elena's autopsy revealed evidence of necrophilia.

What's haunting about von Cosel's story is the motivation behind his necrophilia, which by his account was an obsessive love of one woman. That's very different from the sadistic desire for domination that serial-killing necrophiles have described. And interestingly, people at the time viewed von Cosel's crime in the same rose-colored light. The newspapers of the day generally portrayed him as a harmless eccentric felled by his undying love of Elena. The local paper, the *Key West Citizen,* referred to him as "one of the truest of romanticists of all time." After his arrest, two local citizens posted $1,000 bond for his release from jail, telling the paper, "We who know him think he should be freed of all charges." And his peers agreed: People wrote in from around the country to support him.

As for his beloved, Elena's remains were put on public display after being discovered in von Cosel's home. The spectacle drew more than 6,000 curious visitors, including schoolchildren who were let out of class for the event. Clearly, people weren't too horrified to take a look, though the sexual part of the story wasn't public at the time. (Perhaps they would have found the whole thing a bit less enchanting if it had been.)

Whether you see him as a hopeless romantic or a creep, von Cosel's case is a rare glimpse into the mind of a necrophiliac who was not a killer. There are so few reported cases that research on their psychology is scant. But existing studies suggest they are often men who fear rejection and develop fantasies involving corpses, sometimes after being exposed to dead bodies. They begin to focus their desires and fears onto a cadaver, which can never resist or reject them. In a way, their inappropriate sexual desire shares a bit of similarity with the misplaced lust of a crow—albeit with more complex psychology.

Psychologists who have studied necrophilia say that the majority of necrophiliacs, like von Cosel, aren't killers. In 1989, psychiatrists Jonathan Rosman and Phillip Resnick reviewed 112 cases and concluded that "although necrophiles have been considered 'crazy' because of the bizarre nature of their acts, only 11 percent of the true necrophiles in the sample were psychotic." Of the 54 people they considered to have "true" necrophilia—defined as a persistent sexual attraction to corpses—28 percent had committed homicide. Ninety-two percent of those with necrophilia in the study were male; most were of normal IQ and had jobs (often in hospitals, cemeteries, or morgues). A quarter were married.

None of this is to excuse necrophilia, but rather to try to understand it. Sex with dead bodies is taboo for good reason: At the very least, it's a violation of the golden rule (because chances are most people would not like this particular act to be "done unto them").

Necrophilia is treated as a crime in many countries and most U.S. states—but not all: There's no federal law against it, and a number of states lack specific laws for prosecuting it. As social scientist John Troyer of Ohio State University has pointed out, only four states explicitly ban necrophilia, whereas others use vague wording about "crimes against decency" or abuse of a corpse. So why wouldn't the practice be more clearly defined as illegal, given the public's near-unanimous horror of it?

The answer may lie in our own discomfort. As Troyer observes, to legislate necrophilia, we'd have to acknowledge its existence and debate the merits of a law publicly. This, in turn, raises questions most of us would rather not ponder, including whether a dead body is still a person (under the law, a corpse generally does not have personhood status, so rape laws don't apply). And if it's not a person but an object, who "owns" it?

As with most taboos, though, there are consequences to ignoring the situation. Wisconsin prosecutors found that out the hard way in 2006, when attempted sexual assault charges were dismissed against three young men who dug up the grave of a recently deceased young woman with the intent, prosecutors charged, of having sex with her. A judge ruled that Wisconsin law did not specifically outlaw necrophilia. The case went all the way to the state supreme court, which finally determined in 2008 that the dead can qualify as victims of sexual assault.* Other states have also had to determine the law on the fly. As of this writing, Kansas, Louisiana, Nebraska, New Mexico, North Carolina, Oklahoma, and Vermont have no law against necrophilia.

Although we may not like to admit it, humans have a lot in common with the rest of the murderous, cannibalistic, corpse-bonking animal kingdom. And that's fine. After all, it's not our animal natures, but what we decide to make taboo—or not—that sets humans apart. That's why it's good to reexamine those taboos from time to time, and to ensure they're making our world a better place to live. Taboos should help protect us from the worst we can do to ourselves and others. In that way, at least, they can set us free.

....

* The last of the three men was sentenced in 2010. Two got prison time for attempted third-degree sexual assault, while the third served 60 days in jail for theft and criminal damage to a cemetery.

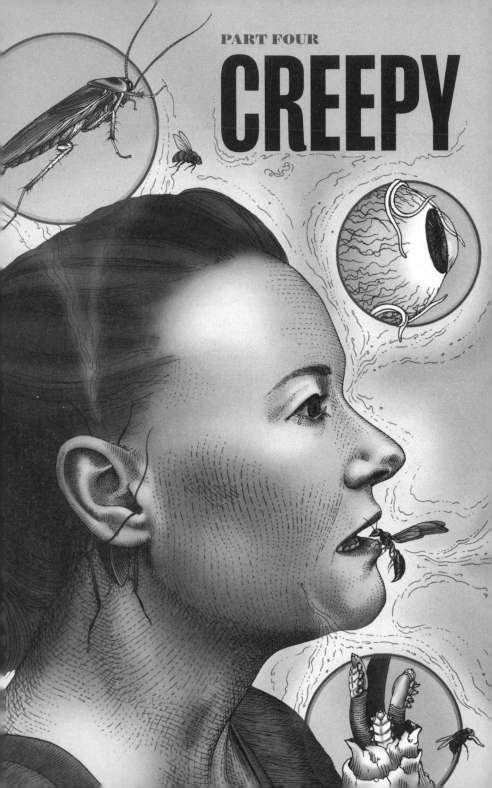

PART FOUR

CREEPY

CRAWLIES

BUG OFF!

Why do certain species give us the heebie-jeebies?

A woman I interviewed recently about insect infestations asked me a question that gave me pause: "So are *you* afraid of insects?"

Of course, I'm not supposed to be. I'm trained as a biologist and a science writer. Insects are some of my most fascinating subjects, from their startling colors and shapes to their brilliant behaviors and evolution. (Just watch a video of the mating dance of peacock spiders and tell me you aren't in love.) And yet.

To be honest, I told the woman, some of them freak me out a little. I appreciate the resilience of the American cockroach, but every time one skitters by, I remember the time a huge specimen managed to get inside my bathrobe. It sputtered against my belly, trying to fly away, and I essentially levitated out of the robe and streaked naked across my apartment. I may appreciate nature in all its gory glory—but I'm not immune to the fear and disgust most people feel for its creepier, crawlier members.

Certain animals are almost universally reviled, and they tend to be the ones that impose themselves on us in the most intimate ways, like the roach in my robe. This includes the mites that live in our hair follicles and lice that have literally ridden our bodies across the globe, evolving so closely with humans that they've specialized into three genetically distinct types: head, body, and pubic lice. Then there are the pests that invade our homes, like cockroaches, flies, and rats, and the ones we're flat-out afraid of—the ones that sting us, parasitize us, or otherwise terrorize us with their horrifying powers.

They give me the heebie-jeebies, too. But I'd like to tell you two stories about my own close encounters with creepy crawlies that shifted my perspective a bit. The first involves flies; the second, rats. From the human perspective, these are two of the most lowly and reviled animals on Earth: Norway rats inhabit homes and sewers, and carrion flies swarm all over dead things.

First, some background on how my encounter with both rats and flies began. It all started with a gnawing sound in the ceiling above my basement. The steady *crunch-crunch* of rat teeth was just a nuisance at first. But then the entire basement began to smell of rat urine, which turns out to smell a lot like people urine. So I called an exterminator, who placed traps that the rats ignored completely. Finally, in desperation, I pushed little green blocks of rat poison into the ceiling space behind the recessed lights. I considered the fact that I'd probably end up with at least one dead rat trapped in the ceiling space, but decided that suffering through the smell of a dead rat or two would be better than allowing them to multiply.

It seemed to work. The gnawing stopped, the basement was stinky for a while, and once the smell subsided, I figured the worst was past. I was wrong. Two weeks later, I came home from a trip and opened the door to Flymageddon. Dozens of giant black insects buzzed around the kitchen like flying bookends, *thunk*ing into me as I made my way in. I realized immediately what had happened: Flies had found their way into the basement ceiling space and laid their eggs on the dead rats, and those eggs had become maggots. While I was out of town, they had grown up to become the swarm of flies now filling my basement. A door in the kitchen led to the basement, and I had closed it before I left for my trip. Now, standing beside that door, I could hear the hum on the other side.

I needed a weapon, fast. Years ago, my Uncle Rocky and Aunt Martha, who live near Abilene, gave me a gag gift in the form of a three-foot-long, turquoise Texas-size fly swatter. Turns out, it was the best present ever.

So there I was. I gripped the Texas Fly Swatter like a baseball bat and slowly opened the door. My pulse was pounding. I flipped on the light and saw thousands of big dark flies, each the size of a dime, peppering the walls and window shades. Flies filled the air and bumped against the ceiling with little buzzing thuds. Suddenly a squadron broke ranks and rushed straight up the basement stairs at me. I made a noise like a creaky hinge and slammed the door.

The flies' numbers were beyond anything I could have imagined. I needed a plan—and a partner. I was home alone, but that didn't stop me from dragging my husband, Jay, into the scene from 500 miles away. I called him on speakerphone blubbering about flies.

The great thing about being married is that you can take turns being brave, and when one of you is freaking out and ready to burn down your house, the other can spring into action. And even from two states away, Jay sprung. "Go downstairs and open all the windows to let them out," he instructed. I politely declined, as in, "NO NO NO NO! I'm not going down there!"

"OK," he conceded. Then, being biologists, we hatched a plan that would take advantage of fly physiology. "Turn off all the lights inside, then go outside and turn on the car's headlights. In fact, put the brights on. Then, open the basement door." Flies, of course, are drawn to light. It's not entirely clear why, but it's the reason you'll often find flies clustered on windows (at least at my house, you will).

It sounded like a plan that might work. So I carefully unlocked the basement door from outside, pushed the door open, and ran for the safety of the car. "Don't fall and hurt yourself running from flies!" Jay yelled, still on speakerphone. "They can't hurt you!" At this point, he was picturing me laid up with a broken leg, a victim of my own horror of animals that don't even have mouths with which to inflict bites.

I knew they were harmless, but that wasn't the point. When it comes to a swarm, it's not about logic. There's something in the primitive part of the brain that screams, "Run!"

So this is how I found myself sitting in my car at 10:30 p.m., watching flies meander out the door and trying to decide how long I could run the brights before the battery died. As if on cue, my mother called. Hoping to help, she looked up flies in the encyclopedia and reported that the pupal stage lasts two weeks. (My mother does not use the internet much.) Her book didn't say how long the adults live. In any case, waiting for them to die inside my house was not an appealing option.

But once I'd drawn out as many as possible, I had nothing to do but wait. And they kept on coming, with more hatching over the course of several weeks. I would discover that all were varieties of carrion flies, which include blowflies and flesh flies. Every morning, I vacuumed up the night's casualties, and every evening I came home to more. One day, I arrived at work, dropped my purse on my desk, and a giant fly flew out.

I thought talking to a forensic entomologist might help me appreciate my new housemates. Sibyl Bucheli studies insects at Sam Houston State University (home to a great criminal justice program) in Huntsville, Texas (home to the busiest execution chamber in the United States). I knew I'd like her when her email arrived with a photo of her wearing a Wonder Woman tiara.

Bucheli told me about the first recorded use of forensic entomology, in the 1300s. It involved carrion flies (possibly one of the species zipping around my head as we spoke). The Chinese lawyer Sung Tzu was investigating a stabbing in a rice field and asked all the workers to lay out their sickles. Blowflies immediately landed on just one, even though it had been wiped clean, and Sung deduced that the sickle bore traces of blood.

As for my flies, Bucheli informed me that I was probably on the second generation, at least. The flies had been multiplying, babies growing up and having babies of their own. I suppose it would be sweet, if the family home being handed down weren't a dead rat. What's more, she gave me bad news about the yellow-orange spots all over my windows. "That's fly poop," she said. "Sorry."

Still, the conversation made me feel a little better about living with a swarm for a couple months. For one thing, Bucheli is fearless when it comes to flies, and it made me want to grow up to be just like her. "I feel calm if I'm in a place with a million flies," she said. "But if I'm in a city with a million people around me, that freaks me out . . . I understand the flies better."

I get it. The flies in my house were just living their lives: eating, mating, pooping, laying eggs. They weren't out to get me or anyone else. "The whole six-legs, four-wings thing is beauty to me," Bucheli said. I can see what she means. Carrion flies are underappreciated as nature's recyclers. They can smell blood or decay miles away, and they're incredibly efficient at their work. Their eggs can hatch into maggots within hours, and in hot weather, they can reduce a body to bone in just two weeks. It's impressive. And yet.

I'm not going to say that I'd be perfectly comfortable sitting in the midst of a blowfly swarm like Bucheli; that would be overly ambitious. But I can say that I'm less freaked out by blowflies and their ilk, having learned more about them during the many weeks it took for them to finally die out.

And that's my hope for you, too. In this chapter, you'll find tales of the animals we shun: the boorish houseguests that munch on our leftovers, the intrusive hangers-on that don't know how to respect our personal space, and the outright nasties that have evolved wicked defenses. But instead of a horror show, let's think of these creepy-crawly tales as a display of nature's ingenuity. These are some of the toughest survivors in the world, honed to take advantage of whatever scraps life offers.

On that note, it's only appropriate next to delve further into the life of the ultimate urban scrapper, and the bane of my D.C. home: the rat.

RAT RACE

Why there's no stopping them

They eat our food. They furnish their nests with our
detritus. They chew through our sheet metal, our lead
pipes and our concrete. They outsmart us at every turn.
They are our shadow, our enemy, our next door neighbor.
—"Rat City!" *Spy* magazine, 1988

You have to think like the rat," my new friend Gregg told me. At the time, we were pushing his homemade rat detector through a small hole in my basement ceiling. Gregg had bought an endoscope camera online—the kind a doctor might use to hunt for polyps in one's nether regions—and attached it to a bent wire coat hanger. The camera's images would be displayed on his laptop.

Gregg became obsessed with rats when they took over his girlfriend Anne's house, across the street from mine. Having tracked and conquered her vermin, he was eager to bring his rat-busting skills and tools to my infestation, which had now waxed and waned over several months. Gregg and Anne had heard all about my invasion, followed by stinking rat corpses and the Flymageddon that hatched out of their carcasses. And now, the rats were back.

Gregg showed up on a Sunday afternoon with the endoscope and a two-gallon bleach sprayer and explained my role: Simply turn the endoscope's light up or down on his command as he threaded the coat hanger through ceilings and walls.

In the ceiling space above the basement bathroom, we hit the mother lode: Towering piles of black rat turds appeared on the laptop screen. "Here's your nest," Gregg proclaimed: our first small victory in a long, losing battle.

I had learned a few things about rats by this point: They are creatures of habit. They establish trackways through a house, following the same paths each day: in, out, to food, to nest. And they can, in fact, rise up from the sewers.

This last point became central to my investigation. When my husband, Jay, cut out a section of the bathroom ceiling where Gregg's endoscope had led us, we found that our rat nest was centered around an old drain pipe that, unbeknownst to us, had been cut but never capped during the removal of an upstairs toilet. Dark, oily smudges marked the pipe's rim, where rats had climbed up from the sewers and dropped into the space between the rafters of my basement ceiling, hidden above the drywall.

Upon further research, I found that not only is it pretty easy for a rat to climb up a three-inch toilet drain pipe (most of the time there's not even water in it), but that I lived in a part of D.C. with a combined sewer system, meaning the storm drains on the street and the pipes from the toilets run to the same labyrinth of sewer pipes beneath the city. A combined sewer is one big happy Rat Central Station, a safe network in which they can run from house to house and pop up in any toilet or drain.

Having figured out how our rats were getting in, and assuming that any remaining rats would have been scared away by our noisy labors and hole poking, Jay capped the pipe, and we congratulated ourselves on a mystery solved.

Maybe you can guess where this is going. Would capping a pipe really be enough to keep a rat out? After all, their superpowers are near mythical: They are rumored to be able to swim for three days. Or fit through holes the size of a quarter. They've even been said to have no solid bones, just cartilage (that one's definitely false).

I looked to science for the truth about rat abilities. But I was surprised by the dearth of studies on the biology and behavior of the Norway rat—the common city rat, *Rattus norvegicus*—in the wild (the "wild" in this case being any city on Earth). As it turns out, despite our long human history with lab rats, we know very little about the ones that could be in our homes. "We probably know more about the ecology of polar bears than we do about rats," says Chelsea Himsworth, a veterinary scientist who is studying how rats spread disease in cities as part of the Vancouver Rat Project.

"The interesting thing about Norway rats is they don't exist in the wild," Himsworth told me. Their migrations, meandering through Asia, over continents, and across oceans, are our migrations. They've been in contact with humans for so long that they not only live with us, but they also depend on us almost entirely for food.

And chillingly, they don't stray far from our homes. One of the most important findings of the Vancouver Rat Project is that rats form highly stable family groups and colonies, block by block in a city. And when people break up a colony—say by indiscriminate trapping or poisoning—the remaining rats are forced to move. And that's when they tend to spread disease.

I was, of course, trying not to be indiscriminate. I wanted to kill them all—the whole rat family. My ordeal had turned me, an animal lover who wanted to be either a veterinarian or David Attenborough when I grew up, into a cold-blooded rat killer. I told this to Robert Corrigan, who has been described as the "rat king of New York City." He seems OK with that title, and showed sympathy for my plight. Corrigan has spent his career fighting rats up and down the eastern seaboard, which thanks to its dense population, waterways, and old pipes is pretty much rat heaven.

Corrigan said he agreed with my friend Gregg: To wipe out an infestation, you have to think like a rat. "But I also think it's not difficult to outthink a rat," he explained. Unlike many animals, a rat must have both food and water every single day to survive. No skipping meals. "If it doesn't have food and water, it goes into this kind of 'crazy mode,'" Corrigan said. Rats have a very low tolerance for hunger, so to get rid of them, simply ask where they're getting food and eliminate the source.

But what about my rats? I asked him. How were they getting food? Clearly they were coming up an old toilet pipe from the sewer, and there wasn't any food in my basement ceiling. That's where it got a little ugly. I was right about the combined sewer system, Corrigan said; it does make it easier for rats to get into toilets. As if to make the point, the day after we capped our toilet pipe, a rat popped up in my next-door neighbor's toilet.

Plus, toilet drainage turns out to be a boon for sewer rats. "Lots of food gets flushed," Corrigan pointed out. (This remains hard for me to fathom, but I do recall a landlord once complaining about a tenant who kept flushing chicken bones down the toilet.) "Also, if push comes to shove, human feces and dog feces contain undigested food," he continued.

Let's pause on that for a moment. What Corrigan is saying is that the rats in my basement ceiling were climbing up and down a toilet pipe into the sewer every day, whereupon they ate and quite possibly dragged back up caches of food that may or may not have included human excrement. This was bad news. "That's repulsive to humans, but it's called coprophagy, and it's part of the reason rats are so successful," he said. "They don't turn up their nose at anything that floats by."

So it was smart of us to cap the sewer pipe. But little did we know that when we cut off their entrance and exit to the basement ceiling, at least two more rats would remain trapped there—or that only one would survive. Survivor Rat chewed its way out of the house, leaving in its wake a gnawed-off air conditioner condensation tube that spewed water into the ceiling space. Loser Rat died in quarters unknown, spawning a new flock of flesh flies as it decomposed.

After I bug-bombed my basement to kill the final round of flies that had emerged from the carcass, I reflected upon the devastation that one family of rats had wrought. There were holes punched through the ceiling, plus water damage from the chewed tube; I spent hours scrubbing fly poop off walls in the house before finally giving up and having most of the interior repainted. And then there was the lingering smell.

Yes, I had learned a lot about the biology of flies and rats through this experience, and had come to appreciate their incredible adaptations for survival. Evolution has honed them well. I appreciated them in the way, I imagine, that Napoleon might have appreciated the strength of the British and Prussian armies as they advanced on him at Waterloo. I may have won a few battles, but it's pretty clear who's winning the war.

SMALL, BUT MITEY

What's microscopic, has eight legs, and lives in your face?

At this very moment, hundreds or even thousands of microscopic eight-legged animals are nestled deep in the pores of our faces—my face, your face, your best friend's face, and pretty much every other face you know or love. For the most part, these critters hang around totally unnoticed—even though, in some sense, they're our closest animal companions.

These animals are mites—tiny arachnids, related to spiders and ticks. They're too small to see with the naked eye, and too small to feel even as they move about. Not that they move much: Face mites are the ultimate hermits, living most of their lives head down inside a single pore. In fact, they're so perfectly adapted that their bodies are shaped like the inside of a pore, evolution having long ago reduced them to narrow plugs topped with eight absurdly tiny legs.

Face mites were first discovered in the human ear canal in 1841; soon thereafter, they were found in the eyebrows and eyelashes as well. Since then, we've learned that they live not only among towering forests of brows and lashes, but also in the savannas of short, fine hairs found all over our bodies, save the palms of hands and the bottoms of feet.

Our skin is covered in several million hair follicles; each sits in a pore and is flanked by sebaceous glands that make an oily, waxy secretion called sebum. These oil-producing pores (which are different from the smaller ones through which we sweat) are particularly dense on the face—as are the mites that live in them.

We call them face mites, though you could just as easily think of them as skin mites or people mites.

Perhaps even more surprising, our pores are home to two different species of mites. Both are in the genus *Demodex,* and each has carved out its own niche in the landscape of the human face. One might imagine that some epic battle for follicular dominance would have long ago been fought and won by one or the other species—but no. Even within a pore, the mites have enough variety of habitat for each to claim its own turf.

Demodex brevis is the shorter and stubbier of the two, shaped roughly like the kind of club a cartoon caveman might carry. It prefers to nestle deeply into sebaceous glands, and is also found in our meibomian glands, which line the rim of the eyelid and secrete an oily substance called meibum* that keeps tears from evaporating.

The other mite species is *Demodex folliculorum,* which is longer and skinnier. If you've ever seen a picture of the little moss-dwelling animals known as tardigrades, or water bears, these mites look a lot like them, but with their legs clustered at one end of the body. As its name suggests, *D. folliculorum* hangs out in hair follicles, closer to the skin's surface.

Both species are such homebodies in their respective parts of a pore that scientists have had a hard time observing them, whether in captivity or in the wilds of the human face. Just extracting one is such a chore that until very recently, scientists weren't sure whether everyone had face mites, or just some fraction of people. And studying them outside a pore is even more difficult, as the mites tend to die within hours of being pulled from their homes. (One scientist claimed years ago to have kept them alive in the lab on a complicated arrangement of layered goat and sheep skins, but no one has been able to replicate that since.)

....

* The German physician Heinrich Meibom (1638–1700) is one of the lucky few to have a secretion named after him. Meibum is an oily substance, which is very different biochemically from sebum; it's composed of a complex mix of lipids and more than 90 different proteins. Issues with the meibomian glands are a leading cause of dry eye syndrome.

As a result, we know relatively little about the mites' lives. But biologists are fairly certain of a few things: Face mites are averse to light. They don't have an anus, so they can't poop.* And they spend their entire lives on our skin. But much of what biologists know about other animals remains a mystery when it comes to face mites. We assume that they eat dead skin cells and sebum, but no one knows the details of their diet. We know they have sex, but not exactly how or when. (Presumably they venture out of our pores at night, when it's dark, to do the deed.)

Because these mites are so cryptic, most of us will never even see one. In recent years, biologist Rob Dunn of North Carolina State University has taken an interest in face mites, and his team has made breakthroughs in understanding these little-known creatures. So, of course, I made it a mission to visit Dunn's laboratory in Raleigh in the hopes of not only seeing my own face mites, but of also learning about these strange beasts. As Dunn told me, he got interested in studying face mites precisely because they're so mysterious—how could something actually live on our bodies without being noticed? Setting out, his questions—like mine—were basic: Do we all have face mites? And if so, are they friend or foe?

Megan Thoemmes wraps her long red hair into a bun and puts on gloves before attempting to remove arachnids from my face. Like me, she's steeling herself for what comes next: scraping the sides of my nose and the creases of my nostrils in hope of squeezing some face mites out of my pores. A graduate student of Rob Dunn's, she's a pro at extracting face mites. But she warns me there's a good chance we won't find any this way.

A better way to collect *Demodex,* Thoemmes tells me, is to put a drop of cyanoacrylate glue, otherwise known as superglue, on

....

* Presumably they don't live long enough for this to become a major problem; their entire life cycle is estimated to be just two weeks or so.

a person's face and then stick a glass microscope slide to it. When the glue dries, you peel it off (it's not as painful as it sounds, she claims) and the glue pulls everything out of the pores, including the mites, all stuck together in a pore-shaped clump. The lab's record-setting find is 14 mites in a single pore.

But on this morning, Thoemmes couldn't find any superglue in the lab, so we're counting on the old-fashioned method: scraping out sebum with a stainless steel laboratory spatula. I'm nervous that she won't find any mites, and that I've driven five hours to see nothing more than a close-up of the gunk in my pores. Thoemmes leans in and scrapes, firmly and steadily. A minute later, she shows me the spatula coated with a healthy smear of translucent face oil, then scrapes it onto a microscope slide and drops on a glass cover slip. Under the scope it goes.

It's hard to say how many mites live on a typical face, because they're so challenging to document. But the human face has about 20,000 pores on average, and multiple mites are often found in a single pore, so Thoemmes says it's reasonable to estimate that many of our faces carry thousands of mites.

Their distribution across the body has been even more difficult to pin down. They've been found wherever sebaceous glands are—including the face, chest, back, pubic area, and nipples—but not everyone seems to have them everywhere. On the face, the mites tend to establish territories that vary from person to person; some folks might have more on the forehead, others on the chin. Once the mites find a hospitable region, they tend to establish a neighborhood and stay put, feeling no need to venture beyond their own mite version of Brooklyn or the Upper East Side. One postdoctoral fellow in the lab, Thoemmes tells me, could always pull lots of mites from one side of his own face, but never from the other; that stayed consistent for years.

Thoemmes adjusts the microscope with the adept hand of someone who has done this thousands of times, and I settle in to wait for the results. I'm not waiting long. After just a few seconds, Thoemmes mutters, "I think I found one." She looks again. "Yes, I did!" We both squeal with joy. Even better, my mite is alive. I watch his tiny legs wiggle as he tries to escape the bright light.

After we take photos and video of my prized former face resident, Thoemmes scans the whole slide looking for more. Slowly, she starts counting. "Two, three . . . oh, I think I may have found a *brevis!*" She had, and we weren't done yet. "Four . . . oh my goodness, you *are* mitey!" she says. Then she is quiet for a long moment. "Eight mites," she announces. Six are *Demodex folliculorum*, and two are *D. brevis*. That's a lot, Thoemmes tells me diplomatically. She usually finds one or two in a face scraping, if any at all. I decide to consider myself above average, in a good way.

Thoemmes herself harbors relatively few face mites, perhaps thanks to her porcelain skin, which is tiny-pored and dry compared with my oil slick. In fact, her own dearth of mites almost prevented her from studying them. When she finished her college degree, Dunn offered to bring Thoemmes on as a graduate student, but only if she could find five face mites in a week. Despite an allergy to adhesives, she started sleeping with sticky tape on her face, hoping to draw one out. She eventually found one mite on her own face—a juvenile—and has only found one more on herself since, even using superglue. Dunn hired her anyway.

Eventually, Thoemmes found a better way to find face mites: using their DNA. In a 2014 piece in the journal *PLOS ONE,* Dunn's group published the first solid evidence that face mites are ubiquitous on humans. When they analyzed the DNA in sebum samples, they found face mite DNA in every single person tested over the age of 18 (versus just 14 percent of people with face scraping).

Further DNA research has revealed that face mites have evolved so closely with humans that at least four distinct lineages of mites mirror our own—with European, Asian, Latin American, and African ancestry. One member of Dunn's team, Michelle Trautwein, who's now at the California Academy of Sciences, is continuing to study this global diversity. She has sampled mites on people from more than 90 countries, and hopes to sequence the entire face mite genome, opening up new avenues of research into their evolution. We might also learn how the mites have evolved alongside us, she says, and getting a look at their genes could help us understand their physiology despite the difficulty of growing them in the lab.

Dunn and his team are ushering in a completely new way of thinking about face mites. The scientists who discovered *Demodex* living on humans in the 1800s saw them only as pests or medical problems—and that way of thinking continued for more than a century thereafter. (Their numbers were found to be greater in people with rosacea, a skin condition that causes redness on the face, and because different species of *Demodex* are linked to mange in dogs, dermatologists assumed face mites cause rosacea in humans, too.)

But now our view of face mites is shifting. If virtually everyone has them, either we're all infested, or that's not the right word to describe their presence. Even their link to rosacea might not be what it first appeared. Thoemmes argues that if mites cause that condition, and everyone has face mites, rosacea should be far more common. Instead, it could be the other way around: Rosacea involves inflammation and increased blood flow, possibly creating conditions that are favorable to face mites. In other words, increased face mite populations could be a symptom of rosacea, not a cause.

What's more, as science has come to view the human body as an ecosystem—home to diverse microscopic flora and fauna—it's not clear that *Demodex* mites can be properly considered human parasites, which by definition cause harm to their hosts. At their normal population levels, they might even help us in some ways, like the "good" microbes that live in our guts. Thoemmes thinks that face mites are likely helping to shape our skin's microbiome, the set of bacteria and other microbes that inhabit it. They could be eating harmful bacteria in our pores, along with dead skin and sebum, or secreting antimicrobial compounds. We and our mites might actually be in a symbiotic relationship: We feed them our pore gunk; they help with the housekeeping.

As for finding my own face inhabited by *Demodex* mites, I feel lucky to have seen them, and hope they're up to something good that an intrepid scientist will eventually identify. In the meantime, I'm proud to say I'm mighty mitey.

ROACHES, DEBUNKED

If you can't beat 'em, build 'em

Kaushik Jayaram seems like a nice guy; he has a friendly smile, and his website is topped with the endearing message "Namaste. Welcome to my page!" But when he starts telling me about the research he conducted on cockroaches for his Ph.D. at the University of California, Berkeley, I can't help picturing a miniature torture chamber.

Inside one cylindrical machine, a disc slowly lowered onto roaches' backs to see how far they could be squished; on racetracks outfitted to record their motions, roaches ran following removal of parts of their legs, or crashed headlong into walls.

Jayaram wasn't trying to make roaches suffer. In fact, most of the roaches in his experiments barely seemed to register their ordeals. Instead, he was testing the insects' limits, seeing just how much weight they could bear, how fast they could run without feet—or missing a leg or two—and how quickly they can scale a wall. Ultimately, his ambition is not to find some hidden weakness to exploit, but to understand what makes roaches the ultimate survivors. Instead of eradicating them, he wants to replicate them. He's reinventing the cockroach in mechanical form, incorporating its superpowers into tiny robots.

After all, if you're looking for a robust design, the roaches have it down. Cockroaches are among the oldest insects on Earth; their earliest (and now extinct) relatives date back approximately 300 million years. Millions of years before the dawn of dinosaurs, cockroaches were getting their start, testing their wings as some of the first animals capable of powered flight (as opposed to gliding).

Today, there are about 4,600 known species of cockroaches, the vast majority of which never encounter human beings. Of these, perhaps 30 are considered pests worldwide. The rest live out their lives in forests, deserts, and even among aquatic plants. Their diversity is astonishing: There are bright red roaches with black spots that mimic ladybugs, giant rhinoceros roaches that can measure three inches long, and even iridescent turquoise cockroaches in Australia.

God may have had an inordinate fondness for beetles, but as Richard Schweid notes in his ode *The Cockroach Papers,* the roach got the best design of all. "It is, undeniably, one of the pinnacles of evolution on this planet," he writes. Their basic form has allowed cockroaches to spread to nearly every habitat on the planet and to survive for millions of years, through mass extinctions and the invention of Raid, with few changes—apart from a growing resistance to pesticides.[*]

And that's exactly what makes cockroaches so irritating to humans: They're invincible. They take a beating and keep on skittering. It's also what makes them an engineer's dream.

Have you ever tried to stomp a roach, only to have it skitter away unscathed? Or seen one disappear into an impossibly small crack? That squashability was one of the first superpowers that Jayaram wanted to unpack. And as he quickly learned, roaches have an incredible ability to compress their bodies and spring back unharmed.

His test subject was the American cockroach, or *Periplaneta americana,* also often called water bugs, palmetto bugs, or as

....

[*] In 2019, Purdue scientists reported that German cockroaches are developing resistance to multiple classes of pesticides at once, which will eventually make even mixed insecticides useless against them. The revenge of the roach is nigh upon us.

we knew them in New Orleans, "the big ones" (the little ones being German cockroaches). To test the animals' limits, Jayaram built tiny tunnels and used that roach-squishing machine in his graduate adviser Robert Full's lab in Berkeley.

First they needed to determine how low a roach could limbo. That's where the roach-squishing machine came in, slowly compressing a roach while measuring the forces on its body. They found they could compress a roach to a space a little over an eighth of an inch high (three millimeters)—just a quarter of its typical standing height of about half an inch (12 millimeters)—without any inherent damage.

Next they put the insects through an obstacle course consisting of smaller and smaller crevices. And as they learned, American cockroaches can squeeze through a gap the height of two stacked pennies in about a second, compressing their body by 40 to 60 percent. Not only can they fit through tight spaces by flattening their flexible exoskeleton and splaying their legs to the side, Jayaram and Full found, but they can also keep running nearly as fast while squished, the team reported in 2016 in the *Proceedings of the National Academy of Sciences*. A roach's top speed is about five feet, or 50 body lengths, per second. Scaled up, that's equivalent to a human running more than 200 miles an hour.

Lest you worry about the fate of the roaches put through the squishing machine, Full assured me that none were harmed in the name of science. "We only pushed them to 900 times their body weight," he says. That's like parking a 9,000-pound forklift on top of a 100-pound person, but even that didn't hurt the roaches. In fact, they ran just as quickly afterward.

And run they do. In fact, American cockroaches are some of the fastest-running insects* in the world, based on speed compared with their size. Their versatile legs and the roaches'

* In a 1999 footrace held by the University of Florida's Thomas Merritt, the Australian tiger beetle, *Cicindela hudsoni*, beat out the roach.

agility are key factors. Back in 2002, scientists in Full's lab stuck tiny cannons onto roaches' backs to see how they'd respond when a leg was suddenly knocked off balance by a miniature explosion. The roaches, perhaps not surprisingly, didn't even break stride. It turned out that the springiness of their legs helped them keep their balance against the cannon's recoil.

To trace the origin of the roaches' speed and agility, Jayaram decided to isolate the factors that contributed to them. He began his experiments by removing the roaches' feet, then testing how fast the roaches could run. "Essentially, we did a whole bunch of foot damage—but it turns out that foot damage doesn't make a difference to these guys," he says. He shows me a video comparison; the footless roach seems completely nonplussed. Only on a superslick surface did it lose its footing.

Once he realized that losing a foot didn't make much difference, Jayaram decided to test whether legs were the key to the insects' celebrated speed. With one leg missing, the roaches ran just as fast. Ditto for two legs gone. At three missing legs, they started to slow down, but still ran 70 percent as fast as they had on six. And even missing four legs, the cockroaches' speed was only reduced by 50 percent. "It's really impressive," Jayaram says. A roach's leg, as it turns out, is like a Swiss Army knife, so flexible and multifunctional that the insect can easily survive if it loses a couple.

As agile as they are, cockroaches are not particularly graceful. That becomes clear if you watch a slowed-down video of a roach running, or even performing acrobatic moves like running off a ledge and clinging to its underside. Its movements are better described as "scrabbling," rather than sure-footed or balletic. But they get the job done.

And their acrobatics are fast; a roach running across a floor can transition, apparently seamlessly, to charging vertically up a wall in its path. Jayaram became interested in how this is achieved. At first, he thought roaches might use their sensitive antennae and eyes to detect the wall just before running into it, then quickly process that information and coordinate their movements to rear back and climb onto the wall surface. He was, not for nothing, thinking like a robot programmer. But once again, the roaches surprised him.

Jayaram built a track with a wall at the end, and started filming American cockroaches as they sprinted down the track and up the wall. When he watched the video in slow motion, he saw that the roaches didn't prepare themselves to climb the wall at all. Instead, they just crashed into the wall headfirst. When they hit, they'd just extend their back legs and use their forward momentum to carry them upward. It took them only about 75 milliseconds to make the transition to vertical.

Far from thinking it through, the roaches didn't seem to use their heads as anything but a bumper (though it is a good bumper). If you look at a cockroach in profile, regardless of species, you'll see that its head is permanently bowed like that of a praying mantis (and in fact, mantids and roaches are closely related, both members of a group called the superorder Dictyoptera). When a cockroach hits a wall, its head and springy joints absorb much of the impact.

Other small creatures have this same ability; for instance, mosquitoes survive being pummeled by raindrops larger and heavier than themselves. But above a certain critical mass called the Haldane's limit, which is just over two pounds, animals can't dissipate the energy of their collisions without undergoing what Full and Jayaram called "irreversible plastic deformation." In other words, big animals gets smooshed. It's similar to how a toddler can fall down and bounce right back up, whereas an adult might break a hip: Grown-ups' momentum is much greater at the point they hit the ground.

Roaches are more like toddlers, and are not afraid to smash into things. In fact, more than three-quarters of the time the roaches would seemingly choose to crash headlong into the wall, rather than the alternative of slowing down and tilting their bodies upward. The head-on approach allowed them to run faster, too, up to their maximum speed.

After finishing his Ph.D. and determining the many factors that make roaches invincible, Jayaram moved on to a postdoctoral position at Harvard, where he continued to work on roachy robots that squeeze and scuttle like the real thing. The latest iteration is the cockroach-size HAMR series, for Harvard's Ambulatory MicroRobot, which can run almost as fast as a roach, climb a steep incline, and fall great distances without damage. The robots take inspiration from roaches' exoskeletons, including their jointed legs—of which they have only four.

Full sees roaches and other arthropods—insects, spiders, and the like—as the next big thing in robots inspired by nature. Unlike other soft robots inspired by worms or octopuses, insect-bots with hard exoskeletons and muscles could run, jump, climb, and fly, while remaining flexible and strong. This could be useful for robots built to search for survivors in the debris of destroyed buildings, or for data-gathering robots deployed into areas with challenging terrain.

"We know that cockroaches can go everywhere. They're virtually indestructible," Full says. For roaches, being able to scuttle quickly through small spaces has allowed them to spread into virtually every habitat imaginable and outrun their competition. Other insects probably have their own versions of these super-squishing superpowers, too, he says.

I asked Jayaram—who starts a new faculty position in 2020 at the University of Colorado Boulder—why, out of all the insects that could have inspired a robot, he picked roaches. Part of it was convenience—they're easy to find and to breed in the lab— and part of it was the large body of existing research on their biology. But he also pondered which animal could be considered the most annoying. "If you think about a hardy animal—one that's really hard to stop, kill, or destroy—one of the first things that comes to your head is a cockroach," he explains. That was a plus, he said, because he wanted to build a robot that was small but tough. Apart from that, his feelings for roaches are anything but warm and fuzzy.

"We find them just as disgusting and revolting as everybody else," Full told me. But at least they're revolting in a useful way.

I'VE GOT YOU UNDER MY SKIN

What to do when an insect invades

A lot of videos go viral. But one from 2017 is particularly memorable. In it, a live cockroach is seen wriggling through pink flesh and is eventually pulled live out of a woman in India; the insect had entered her body through her nose while she slept.

First, let's clear one thing up. Reports at the time stated that the roach was pulled from this woman's skull. It *was* technically inside her skull, between her eyes—but if you stick your little finger in your ear, that's technically inside your skull too. More precisely, the roach had crawled into her sinuses, and M. N. Shankar of Stanley Medical College Hospital in Chennai, India, filmed the extraction.

This is the stuff of nightmares: The idea that one of the most reviled animals on Earth might slip inside your face at night. But what kinds of creatures actually climb into people? More important, what parts of the body do they get into?

To begin, some bad news: Roaches are the most common invaders of the human body worldwide. In a period of two years, a South African hospital pulled a total of 24 critters out of people's ears. Ten were German cockroaches, followed by eight flies, three beetles, a tick, an assassin bug, and a badly mangled moth.

And ears appear to be the most common points of entry, according to doctors that frequently have to remove them. Though no one seems to be tracking hard numbers,[*] "it's actually

....

[*]. According to the National Electronic Injury Surveillance System, between 2008 and 2012, an estimated 280,939 emergency department visits in the United States were for

not an uncommon phenomenon to have a cockroach in the ear," says entomologist Coby Schal of North Carolina State University. "The nose is more unusual." (He meant this to sound reassuring, but it wasn't.)

And it gets worse. In 1985, the *New England Journal of Medicine* reported the case of a patient who came to an emergency room with roaches in *both* ears. The doctors involved immediately recognized, as they wrote to the journal's editor, that fate had granted them an unusual opportunity for experimentation. At the time, the most popular method for removing a cockroach from the ear in emergency rooms was to pour mineral oil into the ear canal and then manually remove the roach. But others had suggested spraying lidocaine, a numbing agent, into the ear instead. With roaches in both ears, the doctors could compare the two approaches.

So, "having visions of a medical breakthrough assuredly worthy of subsequent publication in the *Journal,* we placed the time-tested mineral oil in one ear," the team wrote. "The cockroach succumbed after a valiant but futile struggle, but its removal required much dexterity on the part of the house officer." In the patient's other ear, the doctors sprayed lidocaine. "The response was immediate; the roach exited the canal at a convulsive rate of speed and attempted to escape across the floor." At that point, a "fleet-footed intern promptly applied an equally time-tested remedy and killed the creature using the simple crush method." The docs hoped their experiment, though consisting of only one patient and two ears, would be instructive for the speedy extraction of cockroaches.

And surely it has come in handy since, though emergency doctors are still debating the merits of mineral oil versus lidocaine. Each has its drawbacks. The National Center for Emer-

foreign bodies in the ear. Kids ages two to eight years old were the most frequent patients, and jewelry was the most common foreign body in their ears (39 percent). For adults, it was cotton swabs (50 percent), which people shouldn't be sticking in their ears anyway. Unfortunately, because the NEISS database only tracks consumer products, there's no word on how many roaches were among the foreign bodies.

gency Medicine Informatics suggests the mineral oil approach, which suffocates the insect. The problem with this is that it's slow, and means waiting for a roach to die inside your ear before it can be pulled out. Lidocaine can be faster, but it doesn't always kill the intruder; you might end up with a panicked roach scrambling through your ear, an experience the emergency medicine center describes euphemistically as "unpleasant for the patient."

Why so many roaches in ears? "Roaches are searching for food everywhere," Schal says. "And earwax might be appealing to them." Earwax harbors bacteria that produce compounds called volatile fatty acids. Meat also emanates some of these compounds, "so a roach could go in to explore and then get stuck," Schal says. Likewise, nasal secretions might be appealing to a roach hunting for a midnight snack.

But although they might see your ears as snacking territory, roaches aren't parasites. "The roach is not really interested in being on a human, and he wouldn't be if the human was awake," Schal notes. That's why almost all roach invasions happen while the person is asleep.

The insects in question also don't tend to be big. Though the roach in the Indian video seems large, Schal could tell immediately that it was young and likely a nymph, or preadult form, of *Periplaneta*, a group that includes the large American cockroach sometimes found in houses.

Given its small size, Schal says it's plausible that the roach could have gotten pretty far into the sinuses. The nasal cavity and sinuses are larger than you might think, extending between the eyes and into the cheekbones. And because these are air-filled spaces, an insect can survive in there for a while.

How long? "Maybe one of your readers will volunteer to stuff a roach up their nose and see," quips entomologist Gwen Pearson, insect education and outreach coordinator at Purdue University. (We were joking; don't do that.) The point is: No one really knows, but you're usually better off with the insect staying alive until you can get to an emergency room for professional extraction, for unpleasant reasons you'll read soon.

But a roach in the nose is not as bad as it gets. Not nearly. Some leeches, for instance, are known to enter any orifice they can find, including the eyes, vagina, urethra, or rectum. In 2010, scientists described a particularly unnerving leech species in Peru with huge teeth; they dubbed it *Tyrannobdella rex,* making it a *T. rex* in scientific shorthand. So far the leech has been found only in nostrils, which is reportedly quite painful. But because similar species turn up in other orifices, it's probably just a matter of time until we hear of a *T. rex* up someone's bum. (Let's hope there won't be a video.)

Generally speaking, not many species that are large enough to notice will brave the human rectum. However, flies are not picky, and will invade and consume human flesh by laying eggs that hatch into maggots. It's a common enough problem that there's a medical term for maggot infestation: myiasis, reported throughout human history in the eyes, nose, and rectum.

One particularly detailed report in 1783 describes a Jamaican surgeon's attempts to remove maggots from a patient's nose using everything from rum to powdered mercury blown up the man's nostrils. After those efforts failed, the surgeon tried flushing the man's sinuses with a "decoction of tobacco," which was a solution made by mashing and boiling the leaves. It succeeded in "bringing away the insects in great numbers, and in a very weakly state." (One imagines the patient may have been in a very weakly state by this point too.) Ultimately, more than 200 maggots and a transparent substance "near two inches long" and containing three large insects—presumably the original source of the maggots—were removed. The patient was then "able to go to the mountains to recover his strength," the surgeon reported. (He surely needed a vacation after that ordeal.)

Other times, you might not even realize you've been invaded. This was the case with the cockroach that turned up in a 52-year-old American woman's colon during a routine colonoscopy. She had a roach infestation at home, and doctors suspected she somehow swallowed it whole. Endoscopies, they note, have also turned up ants, ladybugs, yellow jackets, and wasps. Ouch.

If at this point you feel the panic mounting, don't worry; if an insect does crawl into your nose or ear, your biggest risk is an infection. I promised to convince you to leave the insect alive, should you require professional extraction, and here's why: The absolute worst thing that can happen if a roach crawls into your nose is that it dies. Bacteria will swiftly begin breaking it down, and then infection could spread from your sinuses to your brain.

As for the prospect of getting an infection from the live roach roaming your sinuses, that's actually not terribly likely. Though people think of roaches as dirty and covered in bacteria, they actually groom themselves constantly, Schal says. Your biggest risk is crushing the roach while trying to remove it, releasing the copious bacteria in its gut. Now *that* can lead to an infection.

Thank goodness, the odds of waking up with an insect inside you are slim in most places. Reports are most common in the tropics, where there are more insects, and in cases of severe infestations in the home.

Lastly, don't believe every horror story you see online. "I see so many fake videos of spiders in people's skin," says Pearson. A bogus report that a spider climbed into a man's appendectomy scar, for instance, made the rounds online. But in fact, spiders don't burrow into wounds, and certainly can't climb around under your skin.

Your best chance of keeping insects out of your body is to eliminate infestations in your house—for instance, by making sure food is put away and secured, and keeping food out of the bedroom.

"The insects are all around us," Pearson says, "and it will all be OK."

THAT'S NOT AN EYELASH

And you thought a roach in your ear sounded bad

Abby Beckley was salmon fishing in Alaska when she felt something in her left eye. "It felt like when an eyelash is poking you," she says. But try as she might, the 26-year-old couldn't find a hair—or anything else—on her cornea. The feeling wouldn't go away, and after about five days, Beckley was frustrated.

"So one morning, I woke up and I thought, If it's the last thing I do, I'm going to get whatever the heck is in my eye out of there," Beckley says. She screwed up her courage, pulled back her eyelid, pinched the inflamed skin underneath, and gave it a yank. When she looked down, she says, "there was a worm on my finger."

Beckley had just become the first person in the world known to be infected with a particular species of eye worm. Known as *Thelazia gulosa*, the parasite had been seen in cattle eyes—a normal pit stop in its life cycle—but never before in a human's.

What's more, Beckley's is only the 11th human case of *Thelazia* eye worms of any species in recorded U.S. history. (The last known case, the researchers noted in the *American Journal of Tropical Medicine and Hygiene,* was reported in 1996.)

But, of course, Beckley didn't know any of that as she stared at the worm on her finger in the summer of 2016. The small, nearly transparent creature wriggled for a few seconds, then died. She had seen similar-looking worms in salmon, so she wondered if she'd somehow accidentally transferred one to her eye. But then, more worms started to appear, and it became clear this was a bigger problem. "I was just pulling them out, so I knew there were a lot," she says.

By the time she made it to a doctor in Ketchikan, Alaska, Beckley had pulled five more worms from her eye. The doctors there were "legitimately freaked out," she says, but didn't know what the worms were or if they were dangerous.

Worried about the proximity of the creepy crawlies to her brain, Beckley decided to return to Portland, where her boyfriend's father, a doctor, prepared the medical staff at Oregon Health and Science University (OHSU) for her arrival.

At the hospital, "they basically rolled out the red carpet," Beckley says. Doctors and interns gathered, hoping to see the rare eye worms. They seemed a bit skeptical at first, she says, and suggested that what had looked like a worm to her was really just mucus.

But Beckley kept insisting she had worms in her eye: "I kept thinking, Show yourselves! You have to show yourselves!" she says. For the next half hour, she sat with hospital staff gathered around her, waiting for a worm to appear.

"I'll never forget when the doctor and the intern saw it wiggle across my eye," Beckley says. "He freaked out and jumped back, and was like, 'Oh my god, I saw it! I just saw it!'"

As for Beckley, "she handled it all with remarkable grace and stride, and is incredibly strong," says Erin Bonura, the infectious disease specialist at OHSU who treated her. Ophthalmologists managed to snag one of the worms; although it broke in half, they sent the pieces to the U.S. Centers for Disease Control and Prevention (commonly known as the CDC).

That worm and several more pulled from Beckley's eye made their way to Richard Bradbury, head of the CDC's Parasitology Reference Diagnostic Laboratory. The nation's primary resource for identifying rare parasites had analyzed nearly 6,700 mystery samples in 2017 alone. "When you don't know what it is, it ends up on our table," Bradbury says.

The scientists found the case fascinating. "All these parasites are rare, and this one is extremely rare," Bradbury says of Beckley's eye worm. Ultimately, he dug up a German research paper from 1928 to finally identify it as *Thelazia gulosa*, making it the third species of *Thelazia* to turn up in a human eye, along with one in Asia and one in California.

The worms are carried by face flies, which feed on the tears of cattle, horses, and dogs; you may have even seen them buzzing around an animal's eyes. If you can get over the horror of their joint existence, they're a fascinating example of parasitic survival.

First of all, Bonura explains, eye worms can't survive without face flies. The worm larvae can mature only inside a face fly's digestive tract and organs, and then they find their way to the fly's mouthparts. When the fly lands on an eyeball and begins to drink tears, the late-stage worm larvae climb out of the fly's proboscis and onto the eye. There, they finish their transformation into adults and produce more larvae, which must get picked up by another face fly—or else the worms face death.

In Beckley's eyes, "there was no way for them to continue their life cycle, so they all just died," Bonura says—or would have died without reproducing if she hadn't pulled them out. It's still a mystery exactly how the worms got into Beckley's eye in the first place, but Bonura suspects it may have happened when she passed through cattle pastures.

In one bit of good news, the worms tend not to burrow into the eyeball itself, but instead take up residence on soft tissue under the eyelids and around the eye socket. Once they get in, though, there aren't many treatment options. Sometimes, antiparasitic drugs are used to kill them, but these can worsen inflammation.

In Beckley's case, the best treatment was for her to gently pull them out, one by one. Over the course of 20 days, she removed 14 worms. Still, the doctors involved in the case agree that this particular species of eye worm isn't a looming public health crisis.

"Do not panic," Bonura says. Not only is it extremely rare for a person to get a face fly in the eye, it's even more rare for the fly to stick around long enough to deposit worm larvae. The best prevention, Bonura adds, is just to shoo flies away. And if one gets in the eye, remove it immediately. "As long as you're doing what you would normally do, it should be fine," she says, because our instinct to keep things out of our eyes usually protects us.

Beckley's worms didn't leave any lasting damage, and she says her vision is fine. A year and a half later, in early 2018, she had trouble even remembering which eye the worms had been in.

And in case you're wondering (like I did), she didn't keep any: "I did not want to spend any more time with those things than I needed to."

WORMS ON THE BRAIN

And you thought worms in your eye sounded bad

In 2010, Sam Ballard was 19 years old when friends at a party dared him to swallow a slug. Within days, the Australian teen felt pain in his legs, then fell into a coma. When he woke up more than a year later, he was paralyzed from the neck down. For more than seven years, he remained severely disabled: a quadriplegic requiring round-the-clock care. And in late 2018, Ballard died. The culprit, doctors said, was a parasite in the slug called a rat lungworm, a tiny translucent parasite that can burrow into the human brain.

I don't know about you, but for me, worms tunneling through someone's brain is about the most horrific and frightening disease I can imagine—because once they're in, you can't just reach in and pull them out. Most people who are infected recover on their own, without treatment, as their immune system kills off the parasites. But an unlucky few, like Ballard, develop a rare disease called eosinophilic meningitis that can cause permanent brain damage.

Ballard's case is one of about 3,000 recorded instances of rat lungworm disease around the world. As the name suggests, a rat lungworm *(Angiostrongylus cantonensis)* spends part of its life in the lungs of rats. It spreads when infected rats cough up baby worms and wind up swallowing some. The worms pass through a rat's gut and are deposited in its poop. A snail or slug eats that poop and picks up worm larvae, which grow for a while inside the new slow-moving host.

To reproduce, a young lungworm must find its way back into a rat, which usually happens when a rat eats an infected snail or slug. Once inside the rat, the worms make their way to its brain to partially mature, then move on to the pulmonary arteries that lead from the heart to the lungs. In this unlikely environment, pummeled by waves of pumping blood, the worms finally mate.

This explains why things can go so badly when a person eats a slug or snail. As in a rat, an ingested lungworm heads straight for the brain. The worms are sometimes able to burrow through a human brain's protective outer barrier, but once inside, they can't get back out. That leaves worms burrowing in the brain, damaging it physically, as well as causing inflammation while the immune system fights back.

When worms die in the brain, the inflammation can be even more intense, which is why doctors rarely treat the infection with drugs that target them; instead, they treat the symptoms and let the body's immune system do its work. It's rare for people infected with the parasite to develop severe meningitis, but it's often deadly when it happens.

Ballard isn't the only person who's been infected with this dastardly parasite on a dare. At least three reported incidents involve boys or young men who were dared to eat a slug or snail. In 1993, an 11-year-old boy in New Orleans was admitted to a local hospital with a headache, stiff neck, vomiting, and mild fever. "The boy admitted that he had, on a dare, eaten a raw snail from the street some weeks earlier," researchers reported in the *New England Journal of Medicine*. Thanks to a healthy immune system, he eventually recovered without treatment.

Almost any species of snail or slug could potentially pass on rat lungworm disease. And because they show no obvious signs of infection, it's impossible to know whether any particular animal is a carrier. "Snails hold a lot of parasites," says Heather Stockdale Walden, a parasitologist at the University of Florida who has documented the spread of rat lungworm in southern Florida. "Parasites want a host that will be eaten—and snails are food for lots of animals, including birds."

What's more, the parasite is now spreading to new places around the globe. Originally from Asia, rat lungworm is now found in Africa, Australia, the Caribbean, and the southern United States. In Hawaii, where the parasite is endemic, state epidemiologist Sarah Park said that as of 2017, there were about 10 documented human cases of rat lungworm a year.

Perhaps the most surprising tale of rat lungworm's spread is in Brazil. At an agricultural business fair in Curitiba in 1988, one item for sale was a kit containing the giant African snail, *Lissachatina fulica,* and instructions for growing it. The big snails are hardier and meatier than the little garden snails usually eaten as escargot—and were a great small-business opportunity, so the sales pitch went. Soon, people across the country had turned their backyards into escargot farms crawling with fist-size snails.

One problem: Not so many Brazilians eat escargot. The cottage industry collapsed, and Brazilians were stuck with enormous snails sliming across their garden paths. They invaded the environment, feeding greater numbers of the rats and snakes that eat them. And, inevitably, the rat lungworm parasite took up residence in these rats and snails. The first human cases of rat lungworm disease in Brazil were reported in 2007; one study reported seven cases from then through 2013.

Pets and other animals can also lap up snails and slugs in their drinking water. In Florida, Walden says, the parasite has turned up in dogs, miniature horses, birds, and various wild animals. It's believed to have killed a white-handed gibbon at Miami's Metro Zoo in 2004; in 2012, a privately held orangutan in the Miami area died after eating infected snails.

Centipedes can apparently carry the rat lungworm larvae, too. In 2018, Chinese scientists reported that a woman and her son had been infected after eating a redheaded variety purchased at a market as traditional medicine.

As rat lungworm reaches new parts of the world, experts say we're the ones who are going to have to adapt. And a good first step is not eating raw gastropods or centipedes. The same goes for frogs, freshwater crabs, and shrimp if served raw. If you still want to enjoy some escargot (or prawns, crabs, or frog legs), that's fine—just make sure they're cooked to 165°F for at least 15 seconds. And a word to the wise on escargot: The ones served in snail shells in restaurants usually come out of a can—and that's a good thing in this case. Canned snails have been cleaned and precooked, so once the chef cooks them again in a delicious garlic butter sauce, you should be doubly safe.

If you already think avoiding raw slugs is a no-brainer, though, keep in mind that it's pretty easy to accidentally eat small ones that sometimes cling to fresh vegetables—and it's possible that the parasite could be carried in their slime trails. "Vegetables should be thoroughly washed if eaten raw, and beverage and other liquid containers should be covered to keep snails or slugs out," says Sue Montgomery, leader of the epidemiology team at the CDC's Parasitic Diseases Branch. "Removing snails, slugs, and rats found near houses and gardens should also help reduce risk," she adds. Given the cases in Australia and New Orleans, it's also a good idea to teach your kids not to eat any wild critters—even on a double-dog dare.

And if you decide that rat lungworm disease is a good excuse to kill the slugs that have been munching on your garden plants, there are plenty of ways to get rid of them. One option is to sprinkle salt on them (but not on your plants) to dehydrate them to death. Or you can trap them; premade varieties can be filled with beer, which attracts and then drowns slugs. You can also just go out in the evening when slugs are most active and simply squish them. (Just remember that it's important not to leave dead slugs lying around where a rat, pet, or other wild animal might eat them.)

And don't get smug, Northerners. Other parts of the United States may not stay lungworm free for long: "With increasing temperatures from climate warming and snails moving northward," Walden says, "it's just a matter of time."

THE WORLD'S WORST STING

Two scientists make a personal sacrifice

One day while Justin Schmidt was riding his bicycle, something went terribly wrong. "I was huffing and puffing, so my mouth was open, and this damn honeybee flew right in and stung me on the tongue," he says. He tumbled to the ground, flailing in agony. Later he described the sting as "immediate, noisome, visceral, debilitating. For 10 minutes, life is not worth living."

It wasn't the worst sting one could possibly endure (we'll get to that). Its intensity surprised him though—which is surprising in itself, because Schmidt, a University of Arizona entomologist, has been stung over a thousand times by 78 different species of Hymenoptera, the order of insects that includes bees, wasps, and ants. He's now famous for developing the Schmidt pain scale for stinging insects, a four-point measure with descriptions of agony that read like hoity-toity tasting notes for Scotch. (The red paper wasp, for instance, rates a three, with pain that's "caustic and burning, with a distinctly bitter aftertaste. Like spilling a beaker of hydrochloric acid on a paper cut.")

Normally a honeybee sting is nothing to Schmidt; it rates as a 2. "Boring," he says. But the tongue was a whole different matter; clearly, where you're stung is crucial. So when a Cornell graduate student named Michael Smith contacted him several years ago with a plan to sting himself all over his body to map the pain, Schmidt had some advice: Don't sting yourself in the eye, kid. But otherwise, go for it.

Smith, who studies bee biology, had compared notes with beekeepers. Although they all knew that stings varied, no one had ever systematically measured the pain. Are the levels consistent on different body parts? Why do some places hurt worse? "Someone's got to do it," he says, "so as a scientist you go out there and make it happen. It's curiosity; that's what motivates you."

Pain has always been difficult to study scientifically because it's so hard to quantify. For example, even if we all agree that being jabbed with a needle hurts more than being snapped by a rubber band, we would still have trouble rating exactly how much more it hurts. To get around this problem, scientists have devised various numerical scales for pain, like Schmidt's.

Smith chose a 10-point scale to rate stings across his body. He did all of his experimenting on himself (so at least there were no differences of opinion), and developed a specific, standardized self-stinging regime. He used a sting to the forearm as a standard for comparison, setting it as a 5 on his scale. Each day, he began and ended his stinging session by stinging his forearm, reminding himself what a 5 felt like.

Smith proceeded to sting himself with honeybees multiple times in each of 24 places, from the top of his head to the tip of his middle toe. And he didn't dodge the scary bits: the nipple, scrotum, and penis.

After three full rounds of stinging, Smith found that the two most painful places to be stung are the nostril and the upper lip, followed by the penis shaft. These spots all have thin skin, which may allow the stinger to penetrate more deeply and send venom to the plethora of nerve endings found in these areas. The penis got more attention in press coverage when the study was first published in 2014, but Smith told me "the nostril is really where it's at." In his paper, he reports that stings to the nostril "were especially violent, immediately inducing sneezing, tears, and a copious flow of mucus."

It makes sense that the nose, along with the mouth and eyes, are not only targeted by stinging insects, but are especially important to protect, because they're crucial to breathing and vision. "Pain isn't for nothing," Smith says; it motivates us to protect our vital functions.

Michael Smith's scale of the worst places to be stung, combined with Justin Schmidt's pain index by insect, provides a compelling analysis of worst-case scenarios for stings. And that begs another question: What's the apotheosis of insect-induced agony? Both scientists agreed that a bullet ant to the nostril would probably top the chart for intensity. Schmidt described the sting to be "like walking over flaming charcoal with a three-inch nail embedded in your heel."

But Schmidt says one thing might be even more horrific. "A warrior wasp to the nose or lip would be up there," he observes—especially because intense swelling could last for days. "You'd get inflamed and red, and you get the misery down the line."

The bullet ant, by contrast, doesn't cause much swelling, or leave much mark at all. "It's almost disappointing to go through that and be rendered a babbling idiot, but not even have a big red spot to show people," Schmidt says. "They take away even that satisfaction."

Indeed, bullet ants use a particularly nasty molecule to inflict pain: poneratoxin, a small peptide that essentially holds open the molecular doors that would normally turn off the pain signals in nerve cells. Instead, these neurons carry an incessant pain message for hours on end. This is just one of the many ways that stinging insects have evolved to use venom, and it's one of the best for causing pure pain.

And that's the bullet ant's goal. Unlike some snakes and other animals that inject venoms meant to kill prey outright, the bullet ant stings purely in self-defense—and it works. No creature that has messed with a bullet ant once is likely to repeat the mistake.

While doing research in Costa Rica rainforests more than a decade ago, I still remember how carefully I would look *exactly* where I was going. The graduate students in my lab had been awestruck when, before my time there, a bullet ant had stung our intrepid adviser. Alan *cried,* I recall a fellow student telling me with awe. I couldn't imagine what it would take to make Alan cry, and I didn't want to find out.

Despite the pain, Smith and Schmidt say it's been worth it. "I'm living the dream," Smith says. "I'm working with bees." As for Schmidt, "The relationship between stinging insects and humans is really about us," he says. "It's psychological warfare—and they're winning."

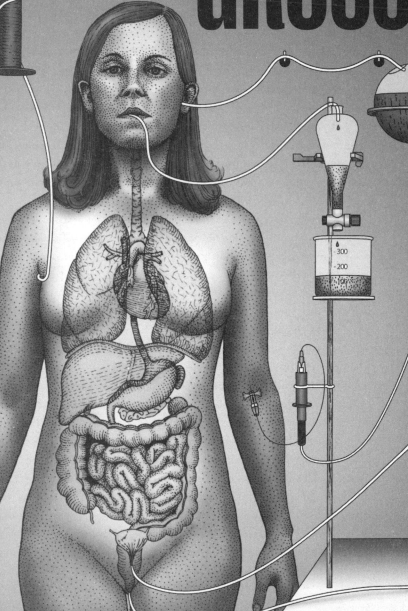

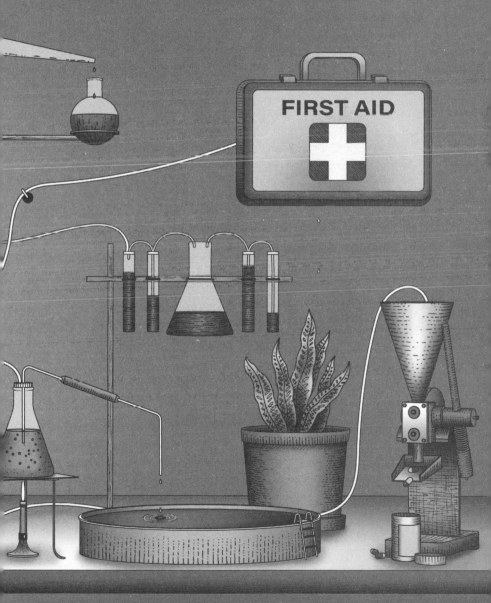

TO SECRETE OR EXCRETE

*Slipping and sliding through
our body's unappreciated creations*

Of all the substances our bodies produce—the mucus, waxes, oils, and fluids—it's hard to say which is the greatest social outcast. Certainly feces comes to mind as the excretion we're most disgusted by, and that we prefer to deal with only in private. Then there's menstrual blood, which carries its own special taboos. Or saliva—public expectoration is outlawed in many places, and spitting on someone is one of the best ways to start a fight. And let's not forget the almost universally reviled earwax.

But as you'll see in the coming pages, even our most embarrassing body fluids (and semisolids) deserve a closer look. In some cases, they're unsung heroes of our anatomy—like feces, which is now being used to cure deadly infections. Other times, they play the villains— for example, our pets' saliva, which can carry a surprising variety of dangerous pathogens. And in yet other cases, they're just misunderstood. It seems intuitive, for instance, that sweat might carry noxious substances out of our bodies—but science suggests that we've oversold that ability.

Still, all this gross anatomy has one thing in common, and it's often the reason we devote more money to getting rid of our secretions than to studying them. Whether urine or feces, blood or sweat, spit or earwax, body fluids disgust us—and thus we avoid them. Case in point: If we can't prevent a drop of sweat from slipping out, we mask its odor, rolling on not just an antiperspirant, but a deodorant too. Collectively, we spend billions hiding our excretions (in 2019, the global market in antiperspirants and deodorants alone was approximately $75 billion).

Likewise, we often shun the parts of the body that produce and carry such substances. People tend to feel the same way about their body's canals and outlets as they do about our cities' sewer systems: There's no reason to think about them as long as they don't clog. And when we do think about them—the orifices and moist crevices—they're the most taboo bits of our anatomy. Yet behind closed bathroom doors, we're plenty engaged with our secretions and excretions. We're waging a quiet war with them.

A word here on the difference between the two. As commonly used in medicine, a secretion is moved from place to place within the body; so for example, a gland *secretes* a hormone into the bloodstream. An excretion is something that's moved from the inside of the body to the outside. Tears, sweat, and feces are all excretions. Earwax is a bit of a sticky case because it's secreted into the ear canal by specialized sweat and sebaceous glands but eventually does leave the body. (So one could argue that your ears are excreting, just very slowly.)

Whatever you call them, these most intimate substances are more important to our health and comfort than we might think. One case in point is saliva, which turns out to hold a host of substances that can give early warnings of disease—from viral illnesses like measles and mumps to cystic fibrosis, which elevates levels of calcium and sodium in our spit. Even when it's just sitting inside our mouths, saliva does more than merely lubricate our food and start the digestion process. It also plays a crucial role in our sense of taste, helping flavor molecules in food reach our taste buds and bathing our tongues in compounds that keep our sense of taste sharp by promoting the rapid turnover of cells.

Even the mucus that lubricates the rectum—a body fluid so unassuming that most of us have never even noticed its existence—turns out to be important if you bother to study it, as one daring group of scientists did. In the chemistry journal *Soft Matter*, a team led by engineer David Hu of Georgia Institute of Technology reported in 2017 that for mammals, the act of defecation takes about 12 seconds—plus or minus seven seconds—regardless of species. The size of the animal and the size of its poop don't matter; cats, dogs, gorillas, and elephants all take about 12 seconds to go.

And why is this? It's all thanks to rectal mucus, the scientists discovered. "Feces slide along the large intestine by a layer of mucus, similar to a sled sliding down a chute," they wrote. The larger the animal, the thicker that mucus layer is, enabling the ejection of a greater amount of poop in the same amount of time. That's interesting in a scatological way, but could also be important to our health: Constipation happens when the intestines do not have enough mucus. By mathematically modeling the flow of excrement, Hu's team found that defecation time can be a useful red flag for diagnosing digestive illnesses—if it takes even a little longer or less time than usual to go, something's off.

But modern medicine has been slow to discover these and other wonders of excretions—perhaps because like the rest of us, most scientists don't particularly want to spend their days immersed in them.

The ancient Greeks and Romans weren't so squeamish. They developed an entire system of medicine built around body fluids. For 2,000 years, Western medicine was dominated by the concept of the four humors: yellow bile, black bile, blood, and phlegm.* A good balance of these juices was considered the key to health and happiness: Too much black bile would make you melancholic, or depressed; too much yellow bile could make you

---

* The humors don't entirely line up with body fluids as we know them today. For instance, modern medicine has only one kind of bile, and it's greenish brown. These ancient categories may have come from observing the colors of blood as it settled and separated into layers of yellow (serum), whitish (white blood cells), red (red blood cells), and black (clotted blood).

fevered or aggressive. The treatment for an imbalance often involved an attempt to drain off whatever was in excess, leading to the popularity of bloodletting for all manner of ills.

Traditional Chinese medicine also emphasizes bodily fluids and ranks them among the body's most important substances. Complementing the vital force, known as qi (or "chi"), the vital substances (called "jin ye") include blood and the fluids that nourish and lubricate the body, including mucus, semen, saliva, sweat, tears, and so on. In this system, as with the humors, too much or too little of any can lead to illness.

If we take a bit of inspiration from the ancients and look more closely at our secretions and excretions, as some brave scientists are doing, we might be surprised what we find. As the following stories reveal, we still have much to learn. For one thing, we've developed many diehard myths and entrenched misinformation over the years. I don't know why, for example, people want to believe that urine is sterile and is therefore an effective rinse agent for wounds—but that claim has been made again and again, and probably will continue to be despite new science showing that virtually no part of our body is actually sterile. And blood-letting has persisted for thousands of years, despite the fact that *you need your blood to live.*

So what else might we learn about our "gross" body bits? What surprises might they hold, whether harmful or healthful? Perhaps as science wades in, we'll find out. Or at least, maybe we can appreciate our oils, fluids, and wastes a bit more as they pass through our lives, instead of cringing and scrubbing all traces of them away. In any case, I hope you'll learn something new about your body, and maybe even see your secretions in a whole new way: as liquid assets.

DIGGING FOR GOLD

When it comes to your ears,
step away from the cotton swab

F ew people truly appreciate earwax. The viscous substance, known to scientists as cerumen, is a mix of secretions made by special glands in our ears. And it's special stuff: Not only does it have antimicrobial properties that protect us from infections, but it also sweeps away dirt as it slowly oozes its way out of our auditory canal.

Yet for all its utility, earwax is "a neglected body secretion," laments George Preti, a chemist who studies human odors at the Monell Chemical Senses Center in Philadelphia. The substance has been studied far less than poop, for example, and possibly less than saliva. In fact, earwax doesn't turn up in today's scientific literature as often as nasal secretions do (but who's counting?).

Preti would probably qualify to be a member of an earwax fan club, if one existed. Given that the same few scientists' names tend to show up again and again in earwax research, the annual gala for the hypothetical organization would be a cozy gathering—more of a cocktail party, really. (If you wanted to fill a banquet hall, you'd probably have to invite the earwax-removal crowd.)

Speaking of which, enthusiastic inventors have a long history of coming up with creative ways to dig out earwax, dating back to at least 1897, when doctor Alfred Hinde reported on his creation of an "iron wire, heated at one end, flattened, and then twisted half a revolution" in the prestigious *Journal of the*

American Medical Association. A drawing was provided; the device looked rather painful.

And yet the tradition remains alive. As recently as 2011, an inventor named Steven Burres applied for a patent for "an earwax removal device having varying structures to provide different earwax removal capabilities." He included 24 illustrations of the varying structures, which amounted to sticks covered in all manner of menacing bumps, bristles, and projections, as well as "a screw-like structure to laterally move wax and grind as it rotates."

Another device, from 2012, looks like the head of an electric toothbrush and includes a system to tell operators how deeply they are mining for gold. Even more disturbing, a patent was granted in 1966 for a device that functioned as a toothbrush, tongue scraper, and earwax remover in one. You wouldn't want to use the wrong end.

Yet ironically, given our obsession with excavating earwax, most of us should just leave it alone. Our ears are self-cleaning, like an oven, and earwax is their Easy-Off. It naturally sloughs off and falls out on its own—so unless a doctor says you have excess or impacted earwax,* there's no need to dig it out. And because the wax is produced in the outer third of the ear canal, inserting foreign objects actually just pushes the wax in farther, forming clogs instead of clearing them. But we persist in doing it anyway. (Cotton swabs like Q-tips actually carry an explicit message not to insert them into the ear canal—but that's exactly what most people do with them anyway.)

Similarly, otolaryngologists—ear, nose, and throat doctors— have a special venom for ear candling (which, if you've never seen it, involves lighting a hollow candle and inserting it into your ear, upon which some combination of warmth and suction supposedly

---

* About one in 10 children and one in 20 adults have excess or impacted ear cerumen, according to the American Academy of Otolaryngology-Head and Neck Surgery. In that case, a doctor should remove it.

pulls out wax). Except that it doesn't work. Ear doctors' feelings on this are nicely summed up by the title of the 2004 paper "Ear Candles: A Triumph of Ignorance Over Science." Basically, tests over the years have shown that ear candles create little to no warmth and no suction—not to mention that drips of hot wax can burn your eardrum. "The inescapable conclusion is that ear candles do more harm than good," concludes the paper, published in the *Journal of Laryngology & Otology*.

Whenever anyone has taken the time to examine earwax instead of lighting things on fire to remove it, they've found it surprisingly interesting. Preti, whose area of focus is mostly armpits and mouth smells (already a sign that he's not the squeamish sort), has found hidden treasure in the scents of earwax. In 2014, he and his colleagues reported that different ethnic groups have distinct earwax odors.

People of European or African descent, he found, tend to have earwax that's yellow and sticky. East Asian and Native American people produce dry, white flakes and crumbles instead. And for those with the yellow kind, I hate to break it to you, but your earwax stinks. Among the compounds that contribute to eau de European are two that are said to smell like sweaty socks—2-methylbutyric acid and isovaleric acid—and another is described as smelling like a goat, hexanoic acid.

The difference stems in part from a tiny mutation in the *ABCC11* gene, just one little letter in the genetic alphabet that long ago granted an East Asian population a reprieve from both smelly underarms and sticky, stinky earwax. Thanks to this gene, most East Asians lack a chemical in their armpits that bacteria munch on to make body odor, and have less of the aforementioned stink compounds in their earwax.

Not only can earwax say something about a person's genes, but it also can help to diagnose illness. Already, earwax is known as a signal for two diseases before they can be detected in blood or urine. One is called maple syrup urine disease, which makes urine smell delicious but is actually a dangerous and deadly metabolic condition. The other is alkaptonuria, or black urine disease, a metabolic illness that makes sufferers pass dark urine that turns black after it sits a while, and also makes their earwax turn reddish brown or black. It's turning out that earwax can be analyzed to detect other illnesses as well, including hepatitis, some cancers, and exposure to toxic chemicals.

And the list goes on. The physical properties of earwax turn out to be no less astonishing. A few years ago, graduate student Alexis Noel—who was studying under David Hu of Georgia Tech, the same guy whose team studied the dynamics of defecation—got curious about the physics of earwax after her boyfriend had to get a blob of it removed from his ear. Many experiments later, she discovered that earwax is a non-Newtonian, shear-thinning fluid: a special kind of substance that's a soft solid most of the time but can flow like a liquid when force is applied to it. (This turns out to be a handy trick. I just bought a cell phone case that uses a non-Newtonian fluid on its edge to absorb the shock of a fall, for instance.)

Noel wondered what a non-Newtonian fluid was doing in ears, so she measured the physical properties of earwax in humans, pigs, dogs, and cows. When she filmed inside the human ear canal, she discovered that moving our jaws to talk or eat creates pressure that gets the non-Newtonian wax flowing, helping it carry trapped dirt out of our ears. She also found that as the wax accumulates dirt and debris, it becomes crumbly, as when you add flour to dough. This helps the old, dirt-filled wax fall out of the ear so that fresh wax can take its place without undue buildup.

It's a nifty trick, and Noel even suggests that it might be useful to think about engineering filtration systems to use sticky coatings that mimic earwax. Someday not too far into the future, for example, you might go for a diagnostic earwax scan at the doctor's office, administered while you shiver in your hospital gown from all the cold air blowing out over earwax-inspired air filters.

Or maybe it's simpler than that: Perhaps next time, you'll just think twice before stuffing a cotton swab or a candle into your ear canal. Even that would be a sign of progress. And who knows? Maybe cerumen will eventually get the respect it deserves.

THE FECAL CURE

The promise and perils of swapping poop

Gross as it sounds, the use of feces to treat illness has a long and colorful history. Sometimes, it's literally colorful, as in the case of a traditional Chinese medicine called "yellow soup," a broth made from a healthy person's stool. References to this remedy date back to at least the fourth century A.D., when Chinese scholar Ge Hong wrote about it in a book appropriately titled *Handy Therapy for Emergencies*.

A number of modern scientists have suggested that yellow soup was prescribed over the years to treat diarrhea. This makes Ge's remedy prescient rather than unpleasant, because we now know that feces can be a cure for serious diarrheal disease when "good" gut bacteria from a healthy person is transplanted into the intestinal tract of a sufferer.

Whether or not ancient Chinese physicians recognized how yellow soup worked, it's entirely possible that some of those who braved the broth did see results, assuming that their abdominal woes stemmed from a microbial cause.

Many centuries later, in 1957, a 23-year-old medical technician named Stanley Falkow revived a version of the yellow soup cure in pill form. A nasty strain of *Staphylococcus* was raging through hospital wards, prompting doctors to treat patients with heavy doses of antibiotics before surgery. As Falkow recalled several years before his death in 2018, the antibiotics had wiped out the bacteria living in patients' guts so thoroughly that their stools were odorless. They began to suffer diarrhea, flatulence, and other abdominal ills as their bacteria-free bowels went haywire.

(Today, modern medicine has established that our guts rely on a unique balance of bacteria and microbes—called the microbiome—for proper digestion, and that overusing antibiotics to kill off bad bacteria can cause problems.)

In response to the uncomfortable situation, one of the doctors working with Falkow came up with this solution: He'd ask patients for a stool sample before administering antibiotics, then reintroduce it into the digestive tract after the surgery. Because the feces would contain the patient's normal gut microbes, the theory went, replacing it would reestablish the patient's gut flora. The question was how to get the poop back into the patient.

So Falkow came up with a plan. He scooped the stool into opaque gelatin capsules that were administered to patients along with other medications. Miraculously, it worked; those taking the poop pills fared better than others after surgery, although they didn't know what they were ingesting. Eventually, the hospital's chief administrator confronted Falkow. "Is it true you've been feeding the patients shit?" he asked. It was, and Falkow was fired (although he was rehired days later).

Decades later, Falkow has been proved right; probiotics and fecal transplants are a hot trend in medicine. A simple approach to maintaining our microbiomes seems to be the most promising: Doctors introduce a healthy person's poop, along with its healthy bacterial community, into a sick person's gut. Today, it's called a fecal microbiota transplant, which makes it sound medical and official, but essentially, it's a poop swap.

Just as in Falkow's era, there remains the problem of getting the feces into patients' bodies without totally grossing them out. Doctors today often deliver a fecal slurry through a tube inserted into the nose, down the throat, and into the stomach or small intestine—a top-down approach—or they take the bottom-up route, using a colonoscopy tube or enema. Neither of these approaches is pleasant, but they do allow doctors to get poop to the part of the gut where it's needed (which can differ depending on the illness). There's also the poop-in-a-pill approach, but it has its downsides too—for instance, a patient may have to swallow 30 or more in an hour.

And yet for a practice that involves such unpleasantness and breaks all kinds of taboos, fecal transplants are becoming a surprisingly popular treatment—especially for infections of the bacterium *Clostridium difficile,* often called *C. diff.* This bacterium is normally a background player in the human gut, a teensy component of the billions of bacteria that equal or outnumber our own cells. But if antibiotics or chemotherapy knock out a person's more helpful bacteria, *C. diff* can take over. In the early 2000s, a particularly virulent strain emerged that stubbornly resists antibiotics: one reason why the bacteria now kill 15,000 people a year in the United States alone. Transplants of fecal bacteria from healthy donors can help reset the microbiome and crowd out *C. diff.*

And now, more research has borne out its effectiveness. Poop transplants worked as well as antibiotics for treating infection with *C. diff,* according to a small study published in 2018 in the *New England Journal of Medicine.* The outcome was striking: Five out of nine patients who received feces from a healthy person were cured immediately of their infections, compared with five of 11 who took antibiotics. That particular study included only 20 people, but the results back up what larger studies have demonstrated in people for whom antibiotics had already failed: Fecal transplants work. With one treatment, they cured *C. diff* at a rate of 80 to 85 percent; that jumped to 90 to 95 percent with repeated treatments, according to a 2018 study in the *Journal of Family Practice.* Antibiotics had only a 25 to 27 percent cure rate.

What's more, poop swaps might help people with a host of other gut-derived problems. Recently, there's been research into using fecal transplants for inflammatory bowel ailments like ulcerative colitis, and even obesity. (Some trials in mice found that fecal transplants could make obese mice thin, and vice versa, although results in humans have been mixed.)

And yet it's surprisingly difficult to get your hands on poop as a treatment. So far, the U.S. Food and Drug Administration (FDA) allows fecal transplants only as a last resort when antibiotics have failed. Patients often suffer for months waiting to see whether antibiotics will work (and they often don't, because they are generally the root of the problem in the first place). That has prompted some doctors—including Michael Bretthauer, an author of the study that compared the two approaches—to suggest fecal transplants as a first resort instead of the last.

Doctors have been reluctant to do fecal swaps partly because of squeamishness, Bretthauer says. "Using feces is a little taboo," he told the *New York Times* in 2018. "If you are putting someone else's feces into a patient, there has to be a good reason."

Nevertheless, many patients aren't too disgusted to try it—especially when they're living with pain and diarrhea that can strike at any moment. Some, in desperation, have turned to do-it-yourself poop swaps. All it takes, after all, is a blender, an enema kit, and someone with whom you're really, really comfortable. Easy peasy. Unfortunately, it's also a terrible idea.

"Some [patients] have even approached us for advice about using their pets as donors," two microbiologists reported to the journal *Nature* in 2014. This horrifying fact was cited in the hopes of convincing the FDA to change their regulation of fecal swaps, making it easier for doctors to perform transplants using prescreened stool. The truth is that dog poop isn't going to cure anyone of much, because dogs carry different bacteria than humans—not to mention the fact that a home enema may not even get the microbes far enough into the digestive tract to be useful. (And you really don't want to shoot Fido's feces into your rectum for naught.)

Even if you're using human poop, it can't just be anyone's, because feces can carry pathogens and parasites that cause serious disease. Even donors who appear perfectly healthy might be carrying bacteria or viruses that their particular microbiomes are able to handle, but yours may not. Your genetics, your immune system, your diet and environment: All these create the ecology of your insides, making it hard to predict the outcome if you start introducing foreign invaders willy-nilly. You could end up with a *Game of Thrones* situation in there as they fight it out.

That's where a nonprofit organization called OpenBiome steps in. It's like a blood bank, but for feces—or as I once dubbed it, the Brown Cross.* After a friend had trouble getting a fecal transplant to treat an infection with *C. diff,* graduate students Carolyn Edelstein and Mark B. Smith decided to open a stool bank. The nonprofit offers hospitals fecal samples that have been prescreened to ensure they're free of pathogens and parasites. Since 2012, they've shipped more than 30,000 stools to seven countries. And the OpenBiome folks have high standards; they've bragged that their poop donor program has a lower acceptance rate than Harvard.

OpenBiome and other groups have also developed capsules containing fecal preparations (an Australian group has dubbed theirs a "crapsule") in the hopes that the pills will gain more widespread use as doctors gather more data on their effectiveness. Ultimately, they aspire to develop pills containing a bacterial mix derived from feces (without containing feces itself) that would go a long way toward transforming the current treatment's ick factor.

Given the success of fecal transplants, it seems inevitable that one day they'll be routine. And I predict we'll adopt the crapsule as soon as it's a viable option. Maybe we'll look back on the early days of fecal transplants achieved by actually *putting poop inside someone* with the same horror we now feel for laying on leeches— but we'll be happy to pop a poop "probiotic" with no problem. In the meantime, if you should find yourself in need of a fecal fix-me-up, my takeaway is this: Kids, don't try this at home.

....

* I mostly try to refrain from toilet humor, but in 2014, when I dubbed OpenBiome "the Brown Cross" in a Gory Details post, other media picked up the name. The folks at OpenBiome loved it and started occasionally using it themselves. No word on how the Red Cross feels about this.

PEE IN THE POOL

It's gross, it's real—and it's slightly toxic

Let's just start with the premise that the next pool you dive into will contain pee. Because it probably will. There's also probably going to be a wee bit of fecal matter, a dollop of human sweat, and some guy's skin cells floating around.

Swimming pools are basically huge blue toilet bowls, providing a constant challenge for their owners. I remember a sign at a childhood friend's pool: "Welcome to our ool. Notice there's no 'p' in it. Let's keep it that way!" Of course, I peed in it anyway. Then my friend told me about the dreaded pool chemical that would turn pee red, surrounding the urinator in a cloud of brightly colored shame. That gave me pause. I spent years worrying that pools might contain the substance, dutifully hauling myself out of the water and going through the hassle of peeling off a one-piece bathing suit to use the bathroom.

It was an effective deterrent—and now I'm going to ruin it. There's no such chemical. It's an urban legend, flat out. Parents have been using it for decades, and pool-supply employees get asked for "urine indicator" all the time. Even though it sounds plausible, the chemistry just doesn't work; you'd have to find a color-changing substance specific to urine (ignoring all other organic stuff that ends up in pool water) that could safely be added to pools in large amounts, so that enough molecules would change color to be clearly visible.

Scientists, as it happens, are not hard at work developing such a substance. That's probably for the best, because if it existed, every pool in the world would quickly end up red. But that doesn't mean science isn't on top of the burgeoning pool-pee problem. Indeed, in recent years researchers have set out to answer the question that plagues guilt-ridden urinators: What happens to pee in a pool?

The answer is not pretty. Chemists reported in 2017 that a typical commercial-size pool contains almost 20 gallons of urine, and on average a residential pool might contain about two gallons. They determined this by measuring the amount of the artificial sweetener acesulfame in 31 swimming pools in two Canadian cities. Because the sweetener isn't digestible and appears in nearly everyone's urine, it's a good marker for how much pee is in a pool.

Another study from 2014 heralded even worse news: Experiments showed that urinating into chlorinated water creates a toxic chemical called cyanogen chloride. This substance forms when chlorine from the pool reacts with the nitrogen in urine. It acts like tear gas, roughing up the eyes, nose, and lungs, and is classified as an agent of chemical warfare. (Predictably, the research quickly resulted in headlines like "Why Peeing in the Pool Is Chemical Warfare.")

But in the grand scheme of bad things that can happen in a swimming pool, how bad is this? Do you really need to worry about dangerous, urine-induced chemicals when you take that next dive? In the 2014 study, researchers mixed uric acid (found in urine) with chlorine in the laboratory. In their worst-case scenario—a cocktail of substances mimicking both urine and sweat mixed with high levels of chlorine—researchers found about 30 parts per billion of cyanogen chloride. That's below the World Health Organization guideline of 70 parts per billion as a maximum concentration of cyanogen compounds in drinking water.

Again, that was a maximum in the lab, not in real swimming pool water. In an interesting thought experiment, reporter Casey Johnston at *Ars Technica* calculated how much pee it might take for an Olympic-size pool to produce cyanogen chloride at a level that would quickly cause "coma, convulsions, and death." Her answer: 2,500 parts per billion.

"In the end, we would need a pool that is two parts water to one part chlorine, which would probably burn the eyeballs out of your sockets and make your skin peel away from your bones," Johnston writes. "If you and three million other people could get at this pool and unload your pee into it before your bodies melted, before the crowd crushed you to death, and before you drowned from the massive tidal wave of pee . . . yes, you could feasibly die of cyanogen chloride poisoning."

When I tried to probe deeper into this calculation, the scientists who had conducted the study were confused; their experiment was intended to illuminate the low-level production in a typical pool, rather than calculate the levels needed to kill an entire pool party. It's probably fair to assume that it would take a heck of a lot of urine to turn a pool so toxic that it would kill you outright. But the researchers point out that there are still health concerns, even for the much smaller amounts generated in an actual pool. Truth be told, it would actually be quite easy to eliminate essentially all cyanogen chloride: Just don't pee in the pool.

Tucked in the paper's supplemental material are calculations the researchers did for a small residential swimming pool used by 20 people. If you assume that only some of them pee in the pool and average out the urine for all 20 people to an average of 50 milliliters per swimmer, or about enough to fill a shot glass, the pool would contain only about 12 micrograms per liter of cyanogen chloride. That's 12 parts per billion—not much, but more chemical warfare agent than you might ideally prefer with your backyard recreation.

Interestingly, the combo of pee and chlorine also produces trichloramine, another lung irritant, and just a touch of chloroform. In recent years, scientists have suggested that trichloramine may be among the reasons that childhood asthma has become more common. In a 2018 study, researchers in Sweden measured levels of trichloramine at swimming pools children use; they found that exposure to the chemical was associated with a higher risk of developing the condition.

This doesn't mean we should all immediately yank our kids out of the pool. Instead, researchers suggest commonsense precaution. If your child has asthma, you might want to monitor his or her condition while swimming—especially in indoor pools, where chlorine-containing gases can build up to higher levels than they do outdoors.

The rest of us will have to keep the problem of toxic pee-induced gases in perspective. If you're weighing the risks of swimming, consider the likelihood of other pool hazards:

Drowning. This is by far your biggest risk, especially if you're a man; in the United States, about 80 percent of people who drown are male. Worldwide, more than 300,000 people die from drowning each year.

Fecal coliform bacteria. Diapers in the pool: enough said.

Other disinfection by-products in pool water. Some of these are known mutagens, meaning they cause mutations, which could lead to cancer. But one review found that even with all these chemicals, swimming pool water is only about as mutagenic as typical drinking water. In other words, this isn't something that needs to keep you up at night.

Your pool mates killing you for peeing in the pool. Could happen.

Personally, knowing that there's some gross stuff in the water isn't going to keep me out of the pool. I do not, however, recommend drinking the water. It is, after all, full of chlorine and pee.

LICKING YOUR WOUNDS

No, a dog's mouth is not cleaner than a person's

In 2007, when Julie McKenna arrived at the hospital in Mildura, Australia, she could barely speak. Her arms and legs were cold and mottled, and her face was turning purple. Doctors quickly determined that McKenna was in septic shock. Bacteria in her bloodstream were attacking her. But even after starting on antibiotics, the purple kept spreading and her organs began to fail. Eventually, parts of her arms and legs began turning black.

After more than two weeks of hospitalization, doctors identified the bacteria in her blood: They were *Capnocytophaga canimorsus,* bugs commonly found in the saliva of healthy dogs and cats. Only then did McKenna remember that she had scalded the top of her left foot in hot water a few weeks before she got sick. It wasn't a bad burn, and she hadn't thought much of it when her fox terrier puppy licked the wound.

Most of us are blissfully unaware of what's swimming around in our own saliva, much less our pets'. And some would rather not know—including those people who love big, wet doggie kisses. I'm not one of those people, and I've found that if I recoil from a dog's attempts to bathe me in saliva, its owner will sometimes say, "Oh, his mouth is cleaner than yours"—as though any logical person would bathe in dog spit every chance they get. Well, I'm here to tell you: Your dog's mouth is not cleaner than mine.

Scientists have recently begun to document all the bacterial species living in dogs' and cats' mouths, and their work is revealing a host of potential pathogens lurking in each slobbery kiss or scratchy lick. In a puppy's mouth, *C. canimorsus* is no big deal: At least a quarter of all dogs and many cats carry these bacteria. Humans normally don't, so once the bacteria got in McKenna's bloodstream, her body struggled to fight off the foreign invader.

Antibiotics eventually turned the tide, but doctors had to amputate McKenna's left leg below the knee, part of her right foot, and every one of her fingers and toes. "It's changed my life in every aspect," she later told ABC News in Australia.

If anyone knows what resides in mouths of dogs and cats, it's Floyd Dewhirst, a bacterial geneticist at the Forsyth Institute and professor of oral medicine at Harvard. Dewhirst pioneered the study of the oral microbiome—all the bacteria that live in the mouth—in humans, dogs, and cats. According to his research, about 400 to 500 bacterial species are common and abundant in the human mouth; around 400 kinds of oral bacteria are found in dogs and almost 200 in cats. Dewhirst expects that more will be found with further study.

Although our skin and immune systems normally stand between us and all those germs, those systems can be breached. About 10 to 15 percent of dog bites become infected, as do up to half of cat bites. Sometimes, the consequences are deadly: In one study, 26 percent of people with confirmed *C. canimorsus* infections died.

One of the main reasons we can get infections from our pets, Dewhirst says, is that our bacterial ecosystems are so different. "If you look at humans and dogs, only about 15 percent of bacteria are the same species," he says. Therefore, our immune systems and native bacteria aren't likely to recognize and fight off foreign microbiota from a dog's mouth. The oral microbiomes of cats and dogs, on the other hand, overlap by about 50 percent.

Part of that difference, Dewhirst says, may stem from what bacteria evolved to eat. Human mouths are dominated by streptococcal bacteria, which are good at eating sugars. "Since cats and dogs normally don't eat too many doughnuts, there's almost no strep," he says.

A single lick can deposit untold millions of these unfamiliar bacteria, which have been detected on human skin hours later. When studying the skin microbiome of humans, Dewhirst and his colleagues were surprised to find several people with patches of skin covered in dog bacteria. "So, if you're licked by a dog, and someone were to take a Q-tip five hours later and rub that spot, they could recover over 50 different species of dog-mouth bacteria," Dewhirst explains.

Weirdly, history is packed with lore suggesting that canine saliva can heal, rather than harm. Dogs supposedly licked wounds at the ancient Greek temple of Asclepius, the god of healing. And there's a long-running but unconfirmed tale that Caesar's army employed wound-licking dogs. But even if it happened, that doesn't mean the practice was a good idea.

It's true that dog and cat mouths have several antibacterial compounds—including small molecules called peptides—that exist in humans' mouths, too. But your pet's tongue is not a magical source of broad-spectrum antibiotics; plenty of harmful bacteria thrive in their mouths as well, despite the presence of these germ-killing substances.

What's more, studies of the antibacterial properties of saliva have been taken out of context, says veterinarian Kathryn Primm, who writes frequently about cats and dogs and hosts the radio show *Nine Lives with Dr. Kat*. One 1990 study found that dogs' saliva had slight antibacterial effects when a mother licks herself and her young, she points out. And a 1997 study in *The Lancet* showed that a nitrogen-containing compound in saliva called nitrite is converted to the antimicrobial substance nitric oxide when saliva is deposited on skin. But both studies involved licking within the same species.

By contrast, after one soldier let a dog lick his wound in 2016, he spent six weeks in a coma while bacteria ate his flesh. That said, such infections are rare; most people with pets will never become infected with *C. canimorsus,* according to the CDC. There's no U.S. estimate of how often these infections happen, because they're not required to be reported to the CDC. But a 2011 study in the Netherlands found that fewer than one in a million people are infected with the bacterium each year. That's a testament to how well our immune systems generally work.

But that statistic raises another question: If your pet's mouth is full of bacteria, how is it actually making itself cleaner when it licks itself? The reason house cats spend so much time grooming is that they retain the instinct as predators. Wildcats use their barbed tongues to "clean blood and such from their fur," Primm says. In large part, they're trying to avoid tipping off prey to their presence. So their purpose is to remove smelly stuff, rather than to wipe out germs. As they groom, cats are coating themselves with the bacteria in their mouths.

Dogs, on the other hand, aren't so picky. "If you didn't clean a dog, it would just be dirty," Primm says. "They aren't stealthy hunter ninjas like cats, so it doesn't matter as much from a survival standpoint." And because they don't groom themselves, they aren't constantly depositing bacteria on their fur.

The good news for owners of fastidious cats is that most feline oral bacteria aren't going to survive indefinitely when deposited on fur. The bad news is that those germs also don't die right away: In 2006, Japanese scientists found nearly a million living bacteria on each gram of cat hair. The scientists also tested how many bacteria were transferred from a cat to a human's presterilized hands by petting the cat for two minutes. The answer should come as slight relief to cat owners: Just 150 or so bacteria made the trip per petting session. As bacteria go, that's not very many.

So petting our pets generally doesn't represent a problem—as long as we keep ourselves clean, too. Primm advises standard hygiene. "After an animal licks your hands, it's a good idea to wash them," she says. The main thing is to make sure those bacteria don't get through the skin; once they do so, bacteria find a moist, happy environment in which to grow, potentially leading to infection.

Licking Your Wounds

As for dogs and their full-face slobberfests, even those are usually not harmful—as long as your immune system is strong and you don't have any wounds on your face or mouth that would allow bacteria into your bloodstream. "Two different dogs licked me in my mouth this week," Primm says.

Nevertheless, keep in mind that babies and the elderly can have weaker immune systems than healthy adults. In one case, parents brought their seven-week-old infant to the hospital with a fever; the soft spot on his skull was bulging. It turned out he had contracted meningitis caused by *Pasteurella multocida,* another pathogen common in cat and dog mouths. His two-year-old brother often let the family's dogs lick his hands, and had a habit of letting his baby brother suck on his little finger.

There's no way to say what the odds of infection are for any one lick. But it's clear that it's not zero—and it isn't true that a dog's (or cat's) mouth is "cleaner" than yours. So wash those hands, and maybe think twice before letting your fur baby plant a big wet one on your face.

TO PEE OR NOT TO PEE?

Hint: Nothing about you is sterile

"If I had to, I'd pee on any one of you." That's a classic line from the television show *Friends* circa 1997, delivered by a well-meaning Joey after Chandler had urinated on Monica's jellyfish sting. The premise was that Joey had seen something on the Discovery Channel about urine relieving jellyfish stings; Chandler ponied up the pee to help Monica out of a painful situation. On the show, the treatment was embarrassing, but it worked. In real life, urine cures are not so reliable.

In fact, pee can actually make jellyfish stings worse. The stinging cells that a jellyfish leaves behind are triggered to release their venom by any change in salinity—so if your urine is too diluted (as in a well-hydrated person), it could unleash more venom. Meanwhile, you're probably already surrounded by a much better salty fluid for rinsing off those stinging cells: seawater.

Jellyfish run-ins aren't the only reason some folks would have you peeing on yourself or your friends. Say you find yourself lying at the bottom of a ravine with a dirt-filled gash in your leg. According to some corners of the internet, you might try peeing on your wound to clean it. Some people would go a step further, recommending drinking urine to prevent dehydration in an emergency, or even sipping it as an everyday wellness tonic.

In 1978, the prime minister of India, Morarji Desai, told reporter Dan Rather during an interview on *60 Minutes* that he drank his own urine every day, and that "drinking urine fights the cause of all diseases." That was pretty much all that a shocked America remembered about the prime minister's visit to our nation, but Desai was describing a practice that some people still subscribe to today.

Now, just to be clear, doctors warn that drinking your own urine is a dangerous proposition, because it's laden with waste products that your kidneys have filtered out for good reason. Even in an emergency—say for someone stranded in the desert without water—the U.S. Army field manual instructs soldiers not to do so. It will only dehydrate you further, because it's only about 95 percent pure water; the other 5 percent is made up of salts, urea, and other wastes. The Army places urine in the same "do not drink" list as seawater, which is only about 3.5 percent salts.

I don't know why we as a species seem to be so obsessed with finding new uses for our urine. But one of the premises underlying these supposed cures is that dousing yourself in it, or drinking it, is safe because urine is sterile, or germ free. Well, wrong again. Urine is not sterile, even if your doctor says so.

For more than 60 years, medical students were taught that any bacteria in urine were there because of infection. But in recent years, scientists have overturned this trope, adding to the growing list of studies revealing that essentially every part of our bodies harbors microbes. "Now that we know they're there, the question is, what are they doing?" asks Evann Hilt, one of the scientists at Loyola University of Chicago who conducted the research. Most likely, she says, "it's like any other niche on our body. You have a good flora that keeps you healthy."

It appears that the urban legend about urine being sterile is rooted in the 1950s, Hilt explains. At that time, epidemiologist Edward Kass was looking for a way to screen patients for urinary tract infections. Kass had developed the midstream urine test (still used when you pee in a cup) and set a numerical cutoff for the number of bacteria in normal urine—no more than 100,000 colony-forming units (cell clusters on a culture dish) per milliliter of urine. A person tests "negative" for bacteria in their urine as long as the number of bacteria grown in a lab dish falls below this threshold. "It appears that the dogma that urine is sterile was an unintended consequence," Hilt says.

So Hilt and her colleagues decided to try a more sensitive technique, reasoning that some bacteria in urine might have evaded notice because they don't grow readily under the conditions of the usual test. They were right. First they used catheters to collect urine directly from the bladders of 84 women, half of whom had overactive bladder syndrome, which causes patients to urinate frequently. Then they put samples in lab dishes and let the urine bacteria grow under friendlier conditions. More than 70 percent of the urine samples contained bacteria that don't normally show up in the standard test. What's more, the mix of bacteria appeared to be linked to bladder problems: Women with overactive bladders had more types of bacteria in their urine, including four species found only in overactive bladder patients.

This finding might provide hope for the 15 percent of women who suffer from overactive bladder; many don't respond to the standard therapy that treats the condition as purely a muscular problem. And understanding that urine isn't sterile also changes the way we think about infection. If the bladder has a normal community of bacteria, we may need to think about its microbiome in terms of "healthy" and "unhealthy" mixes of bacteria, just as we now do for the gut. And in fact, the Loyola team now has some idea of what a healthy mix of bladder bacteria looks like. As they reported in 2018, the microbiome in women's bladders is similar to that of the vagina.

Today, it's not clear which, if any, body parts are actually sterile. Everywhere scientists look, they're finding microbiomes. Evidence now shows that our ovaries and testes may harbor their own bacterial communities, and there's an ongoing debate about whether the placenta is sterile. The womb was long thought germ free, unless something went wrong—but in 2013 Indira Mysorekar, a microbiologist in St. Louis, reported finding bacteria on the baby's side of the placenta. Some evidence also shows babies are born with bacteria already in their guts, which could have traveled there through the placenta.

And what about our brains? Surely the braincase is the last bacteria-free bastion, protected by the blood-brain barrier? Sadly, no. When I asked *Science News* neuroscience reporter Laura Sanders if our brains are sterile, she promptly said, "Oh, no. The brain's full of all kinds of junk." That includes viruses and bacteria.

In 2013, researchers even reported finding bacteria most commonly found in soil in people's brains. (Before making a dirty-mind joke, remember that these are widespread bugs; there's no reason to think soil itself was the source.) The researchers were studying whether people with a compromised immune system from HIV/AIDS might be prone to brain infections. Instead, they found that all the brains they looked at contained bacteria, regardless of HIV status. No one knows how or when the bacteria got in.

So now, some scientists are beginning to study which bacteria are lurking in our brains and what they're up to. Because you can't just go digging around in people's heads while they're alive, researchers mainly rely on brains collected shortly after death, during autopsies.*

In one of these studies, Canadian scientists compared the autopsied brains of people with and without multiple sclerosis. Bacteria in the gut are suspected to play a role in multiple sclerosis by triggering inflammation, and the researchers hypothesized that bacteria in the brain might also be involved. They found enough differences in the brain bacteria of the two groups, as well as links between bacterial proteins and lesions in the brain, to suggest that brain bacteria might play a role in the disease. It's another example of how the more we look, the

....

* This raises the question of whether the bacteria invaded the brain after death. The researchers were careful about contamination and reported that it was unlikely that the bacteria had arrived postmortem. For one thing, they weren't the same kinds of bacteria that were present in surrounding tissues.

more connections we find between our own health and that of our microbes.

And those connections bring us back to our original question: If urine isn't sterile, does that mean you shouldn't pee on a wound? Well, that was probably never a great idea anyway. If you don't have access to clean water, you're generally better off letting your blood flush out a wound, and thereby bathing it in infection-fighting white blood cells.

So if knowing bacteria are in urine helps you talk a well-meaning friend out of peeing on you in an emergency—well, you're welcome.

THE NEED TO BLEED

A brief history of bloodletting as a cure

In the shadow of one of India's largest mosques, the gutters run red with blood. It's a bizarre scene that feels almost medieval—especially if you've never witnessed a modern-day bloodletting.

The method is precise. First, professional bloodletters wrap patients' arms and legs with straps as tourniquets, to control the blood flow. Next, they use razor blades to make tiny pricks in the hands and feet; blood trickles into a concrete trough stained red with the day's work. Meanwhile, the bleeding people look pretty happy. They come to be cured of everything from arthritis to cancer, and pay for the service.

Why hasn't the bloodletting business—which doctors today would classify as quackery of the first order—dried up? Simple: It's long been marketed as a miracle health panacea. Or to put it another way, pain and illness happen "when the blood goes bad," says Muhammad Gayas, who runs his bloodletting business in the garden of Old Delhi's Jama Masjid mosque.

The history of bloodletting began at least 3,000 years ago with the Egyptians. The Greek physician Galen developed his own rationale for the practice in the second century B.C.; his (incorrect) idea was that the liver was responsible for making blood, and sometimes produced an oversupply that had to be let off to keep the body in equilibrium. Equilibrium remained an obsession among many doctors through the 19th century; incredibly, bloodletting was still a recommended treatment for pneumonia in a 1942 medical textbook.

But does bloodletting ever do any good? Well, yes—but rarely and only in specific circumstances. Doctors still use it in cases of polycythemia—an abnormally high red blood cell count—and to treat a hereditary disease called hemochromatosis, which leaves too much iron in the blood. But as a remedy for cancer or the everyday ills of those at the Jama Masjid mosque: No. Although modern medicine has thoroughly debunked most of its supposed curative powers, the practice continues to be used in parts of India, China, and elsewhere as a form of traditional medicine.

Not only are the touted health benefits of bloodletting dubious, but depleting the body's blood supply can also be risky. First, there's the chance of losing too much, which can cause a dangerous drop in blood pressure and even cardiac arrest. The human body contains only nine to 12 pints on average; it takes between four and eight weeks to completely replace a lost pint. Bloodletting in people who are already sick has a much higher chance of causing infection or anemia. Not to mention that in most cases, it just doesn't cure what ails you.

Unfortunately, this was not at all obvious during the Middle Ages. At that time, barbers often doubled as surgeons, putting their razors to work for bloodletting as well as pulling teeth and setting broken bones. Eventually, as organized medical training took hold, barbers were banned from surgical practice. But today, a reminder of that legacy lives on in iconic red-and-white barbershop poles: The red and white represent blood and bandages, while the pole represents the stick that patients would grip to make their veins stand out during the process.

It took some bloodletting disasters in the 18th century to start turning the cultural tide against the once time-honored practice. The prominent doctor Benjamin Rush (a signer of the Declaration of Independence) set off a fury when he began drawing patients' blood during the 1793 yellow fever epidemic in Philadelphia. By all accounts, Rush was a bloodletting fanatic and a real piece of work: "unshakable in his convictions, as well as self-righteous, caustic, satirical, humorless, and polemical," writes his biographer, Robert North.

Rush recommended that, in some cases, up to 80 percent of his patients' blood be drained. During the yellow fever outbreak, North recounts that "so much blood was spilled in [Rush's] front yard that the site became malodorous and buzzed with flies." A later estimate suggested that nearly half of Rush's patients died.

Several years later, on December 14, 1799, Benjamin Rush was embroiled in a legal case concerning his bloodletting practice: Having been accused of killing patients, he sued his accuser. The very same day Rush awaited the verdict in that case, President George Washington lay dying in his bed at Mount Vernon. Washington, recently retired, was suffering from a severe throat infection, and his doctors bled him several times that day, taking an estimated five pints of his blood, nearly half his body's supply. That night, he died, leaving many to blame bloodletting for the death of America's first president.

After such incidents, bloodletting's detractors began to grow, triggering the great bloodletting war of the 1850s (granted, the conflict was fought largely in the pages of medical journals). Pierre Louis, the founder of the practice of medical statistics, began convincing doctors to rely on quantifiable evidence over anecdotal "recovery stories" of patients who had been bled.

One particularly impressive analysis by the doctor John Hughes Bennett showed that among patients with pneumonia, one in three who had been profusely bled died; those Bennett had treated with supportive therapy, such as fluids and nutrients, had all lived. After years of bitter disputes among doctors, the practice began dying out.

That's not to say bloodletting doesn't occasionally pop up in the medical literature. I came across a preliminary study from 2002 suggesting the practice endowed vascular benefits in some with diabetes and abnormally high iron levels. But this is far from a general or accepted treatment for the disease.

Another small study in the journal *BMC Medicine* got a lot of press attention in 2012 for showing that 33 people who were drained of up to a pint of blood had improved cholesterol ratios and blood pressure six weeks later, compared with people who didn't give blood (which the doctors attributed to a reduction of iron levels). But too little iron can also be a problem. What's more, the amount of blood removed in the study was fairly low; a pint is about as much as you'd give when donating blood (which for the record, is a great thing for healthy people to do). More important, the design of the study doesn't rule out a placebo effect, which has certainly contributed to bloodletting's popularity in the past.

One history of bloodletting, published in the *British Journal of Haematology,* refers to stamping out the practice as a triumph of reason and "one of the greatest stories of medical progress." And it was a triumph—but all thanks to the persistence of those who insisted on scientific evidence over anecdotal recoveries, although that was not the mainstream medical thinking of the time.

It's amazing to realize that so many people could be wrong for so long, believing they were healing people even as so many died. And it's a lesson we should all take to heart when faced with any "miracle cure": The plural of anecdote is not data.

THE DETOX MYTH

Can we really sweat out our own toxins?

Sweating used to be taboo. (Remember when women used to assert that they didn't sweat—they glowed?) But peruse any fashion or beauty blog today, and you'll see that sweat is in style—at least as long as you do it at the gym. From infrared saunas to hot yoga, towel-soaking activities are touted not only as relaxation tools, but also as ways to stay healthy by flushing toxins out of our bodies. As the thinking goes, sweat will carry away whatever chemical nastiness we've picked up from our environment.

Too bad you can't sweat away toxins any more than you can sweat actual bullets. We sweat primarily to cool ourselves,* not to excrete waste or toxic substances. That's what our kidneys and liver are for.

Of course, there's often a grain of truth at the heart of a myth, and sweat detox is no exception. Though sweat is made up mostly of water, it can contain trace amounts of hundreds of substances, including some that are toxic.

"You always have to ask how much," says chemist Joe Schwarcz. Plenty of substances are found at very low levels in sweat. But just because something is present doesn't mean there's enough of it to create a health risk, he explains. Schwarcz directs McGill

....

* Humans are almost, but not quite, unique in cooling off through the evaporation of sweat. The only other animal that sweats profusely to cool itself is the horse. (Most mammals pant.) Interestingly, the reason horses' sweat lathers up is that their apocrine glands produce a protein called latherin, which acts as a natural detergent. The foaming action helps sweat pass through a horse's water-resistant coat to the surface, where it can evaporate.

University's Office for Science and Society, which debunks science myths; the group is inundated with questions about medical scams and quackery, including many that promise to "detoxify" the body. "Whenever you see this concept of 'detox' in the popular media, you're usually looking at silliness," Schwarcz says.

For one thing, most "detox" products and diet plans are pretty vague about what toxins exactly we need to rid ourselves of. Pesticides? Metals? Whatever makes up processed cheese? Whatever they are, toxins sound nasty and we want them *out*. And because we can't see them, it's pretty easy to convince people that fasting or drinking green stuff or sweating a lot will do it. (After all, if it's unpleasant, it must be good for you, right?) But when you consider how toxic substances actually accumulate inside us and the body's means of getting rid of them, you'll realize that most detox plans make about as much sense for your health as a tapeworm diet.

To begin with, the type of sweat we produce when we're hot or exercising comes from eccrine glands. You have about three million of them all over your body. Because we make this sweat to cool ourselves, it makes sense that it's more than 99 percent water. Dissolved in that water are small amounts of minerals like sodium and calcium, plus small amounts of various proteins, lactic acid, and urea.

Urea—which is produced in the liver by the breakdown of proteins in food—is a waste product, so it's true that sweating flushes a teensy bit from the body. But the process plays only a minor role in our bodies' waste removal systems; for the most part, your kidneys handle the heavy lifting, and the vast majority of urea leaving the body is in urine. Only if your kidneys are failing does sweat become an important way for your body to get rid of that particular waste product.

As for man-made pollutants, the levels found in sweat are so low that they're essentially meaningless, says Pascal Imbeault, who led a 2018 study calculating these levels. As an exercise physiologist at the University of Ottawa in Canada, Imbeault studies fat—specifically, what happens to the pollutants stored in body fat. Known as persistent organic pollutants, these include pesticides, flame retardants, and polychlorinated biphenyls, or PCBs, which have been banned but are still found in the environment.

These are the kinds of chemicals that many people think of as "toxins" in our food and environment. But as Imbeault notes, even that term isn't quite right: Toxins are harmful substances made by living organisms like plants, animals, or bacteria; manmade substances that are toxic are called toxicants.

Whatever you call them, one reason you won't find high levels of persistent organic pollutants in sweat comes down to basic chemistry, Imbeault continues. Sweat is made mostly of water, and these fat-loving substances don't dissolve well in it. (That's why we say people who don't get along are "like oil and water"; they don't mix.)

Imbeault and his colleagues did some calculations to see how much of these pollutants might be excreted, based on previous measurements of fats and pollutants in sweat. They found that a typical person doing 45 minutes of high-intensity exercise could sweat a total of two liters in a day—normal background perspiration included—and all that sweat would contain less than one-tenth of a nanogram of these pollutants.

To put that in perspective: "The amount in sweat is .02 percent of what you ingest every day on a typical diet," Imbeault says. If you really pushed it on your exercise regime, you might release up to .04 percent of your average daily intake of pollutants. In other words, there's no way you could sweat enough to get rid of even one percent of the tiny amount you'll eat in your food that day.

Keep in mind that the levels of pesticides and other pollutants in most people's bodies are also extremely low to begin with. It's a testament to analytical chemistry that we can detect a compound down to parts per trillion, Joe Schwarcz says. But that doesn't mean it's harming you at that concentration, or that incrementally decreasing it to an even lower level will have any health effect.

But back to that grain of truth: Small amounts of heavy metals like lead and bisphenol A (BPA) contained in plastics do make their way into sweat, because these pollutants dissolve more readily in water than the fat-loving ones do. But again, the amount removed by sweating is relatively low; when someone has been poisoned with high levels of heavy metals, a far more effective treatment is chelation therapy, which uses drugs that bind to heavy metals in the body and are then filtered out by the liver and kidneys. And because far more BPA leaves the body through urine than through sweat, you're more likely to rid yourself of that chemical on the toilet than in the sauna.

Not that you need to start gulping down gallons of water, either. Instead, the best way to reduce your BPA exposure is to avoid eating and drinking out of containers made with it, according to the very practical researchers at the National Institute of Environmental Health Sciences.

Likewise, if you're concerned about pesticides and other pollutants in your food, you're better off avoiding them in the first place than trying to sweat them out later. And to help filter out whatever you do take in, you can keep your kidneys healthy—by reducing or avoiding smoking, high blood pressure, and heavy use of nonsteroidal anti-inflammatory drugs (NSAIDs), which are anti-inflammatory drugs such as ibuprofen. Dehydration also stresses the kidneys, so ironically, sweating heavily without drinking enough water could harm your body's ability to cleanse itself. As much as we'd all love a quick fix, a boring old healthy lifestyle* is still the best we can do.

All the same, none of this has stopped the growing sweat-detox industry. The latest trend is infrared saunas, which use light instead of electric heaters or steam to create heat. Sauna use has been cor-

---

* For more on avoiding pesticides and other toxic substances in food and household products, the same common sense applies. You can look to science-based sources such as the U.S. Centers for Disease Control and Prevention, rather than websites and companies that are trying to sell you something. The nonprofit Environmental Working Group also offers helpful consumer information.

related* with better cardiovascular health—possibly because when we're hot, our heart beats faster, as in moderate exercise. But if you see claims that infrared saunas have special detoxifying properties, think again: No credible science demonstrates that saunas, infrared or otherwise, can cleanse us of toxins.

Sweat therapy can also be dangerous if taken too far. For one thing, most people should not stay in a sauna for more than 10 minutes at a time, according to the American College of Sports Medicine. And like most things, just because a little bit is good doesn't mean more is better. In 2011, a self-help guru in Arizona was convicted on three counts of negligent homicide when three people died after a two-hour sweat-lodge ceremony. The same year, a 35-year-old woman in Quebec died after a detoxification spa treatment plastered her with mud, then wrapped her in plastic and put a cardboard box over her head. She lay under blankets for nine hours, sweating. Hours after the treatment, she was dead from extreme overheating.

"It's the old story of wanting to provide a simple solution to a complex problem," Schwarcz says. "Hope is so precious, but some people use hope for selling crazy stuff to people who are vulnerable." Indeed, a huge wellness industry makes it almost impossible to sort through all the claims for diets, pills, and exercise regimes that promise to clean you out, firm you up, or make your skin glow. Who doesn't want to find that "one weird trick" for losing weight, or not losing your hair, or whatever the clickbait is promising this time?

It's too bad there's no one weird trick for blasting away environmental pollutants. But you can still feel good about going to the gym and working up a sweat—for the exercise if not the detox. The guy using the treadmill after you might not love your sweat. But your heart will thank you.

....

* Note that word "correlated." A 2015 study in the *Journal of the American Medical Association* reported that in Finland, where saunas are said to be as numerous as televisions, men who more frequently used saunas were less likely to die of heart disease. But as the authors state, "further studies are warranted to establish the potential mechanism that links sauna bathing and cardiovascular health."

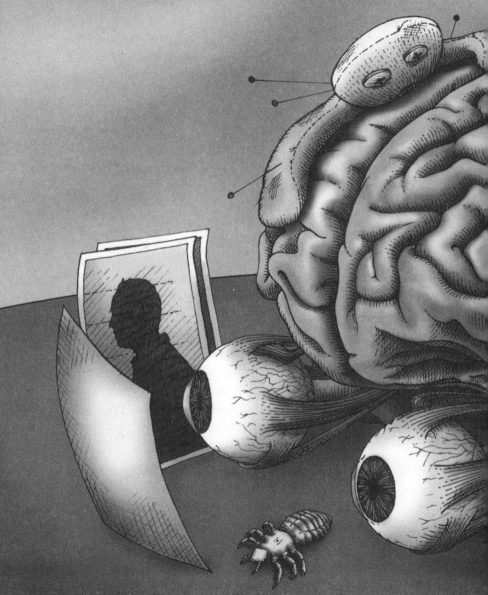

PART SIX

MYSTERIOUS

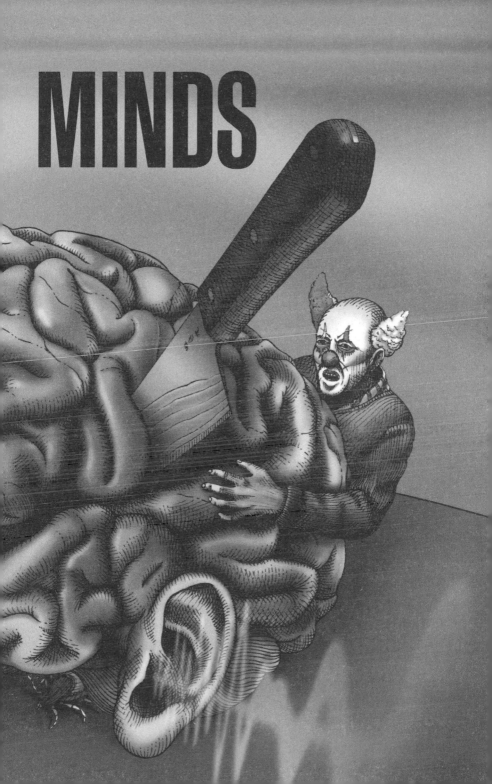

A BUG IN THE PROGRAM

Why our minds play tricks on us

On a steamy July day in Athens, Georgia, an entomologist named Nancy Hinkle is showing me her collection of invisible insects. It's an impressive assortment: A long table is littered with cardboard boxes, mailing tubes, and—most intriguing of all—a large, silver-colored briefcase. All contain samples stockpiled by people who report that their bodies are infested with insects or other tiny creatures.

Inside the mailings are sticky traps that have been set to catch these organisms, as well as cellophane tape that people have applied to their skin in places where they feel the bugs. Some have even sent in the contents of their vacuum cleaner bags—a nightmare to pick through. These samples have been collected in the hope that Hinkle will be able to tell them what species of "bug" is infesting them—and how to get rid of it.

But here's the thing: There aren't any creatures in these samples. They were sent to Hinkle by people suffering from a condition known variously as Ekbom syndrome, delusional parasitosis, or delusions of infestation—the unshakable, but incorrect, belief that small creatures have invaded their bodies. Often, sufferers feel a skin-crawling sensation that they're convinced comes from bugs walking or jumping on them, or even burrowing through their skin. The fact that they can't see the bugs is no dissuasion; they're certain that the creatures are just too small to see.

Sure enough, when I lift the lid on the silver briefcase, it's stuffed with clear plastic bags, some of which appear to be entirely empty. Each is labeled with a sticky note denoting the part of the body or home that one man sampled for insects. One is labeled "top of head," another "head of mattress." Then I spot an empty pill bottle labeled "on penis hole." I show it to Hinkle, and she's nonplussed. She gets things like that all the time, and worse. "My lab members insisted that I get rid of the bag of vomit," she says.

This has put Hinkle and some of her colleagues in an uncomfortable place—and not just because the samples people send are so intimate. Entomologists are experts on insects, not psychology. But because this particular condition revolves around insects, they often find themselves in the awkward position of delivering the news that the insects aren't there. And needless to say, no one wants to hear that the harrowing infestation they've been experiencing is all in their mind.

We think we know our own minds, after all. And how disconcerting must it be to learn that you can't trust what your own brain is telling you? Sometimes delusions are a symptom of a broader illness like schizophrenia. But delusions of infestation are themselves an odd creature. Often this particular misconception occurs on its own, in people who don't have schizophrenia or another serious mental illness, and who have no reason to think the bugs might be a figment of their imagination. And that, to me, is the most frightening part of this assiduously collected display of invisible insects: the idea that *this could be me,* or someone I love.

Clearly, there's something to the old saying that the mind works in mysterious ways—and we'll explore a few of those mysterious workings in the following stories. From delusions of infestation to the psychology of voodoo dolls and why clowns are so creepy, what's become most vivid for me as I've written about the mind is that far more is going on in our brains than we're consciously aware of. And a lot of it runs counter to our sense of self as rational beings in full control of our behavior.

After all, delusions are just one way our minds play tricks on us; all day every day, our brains take small shortcuts that can lead us astray. Here's one classic case: In moments when we can easily find examples of a given topic, we unconsciously come to believe that topic is common. Psychologists call this the "availability heuristic"; it's why people tend to overestimate the likelihood of rare but news-grabbing events, like being killed by a shark. More people actually die of constipation* than shark attacks—but because constipation deaths don't make the evening news, they don't come to mind as easily. Likewise, the availability heuristic ramped up the perceived threat of creepy clowns running amok in 2016, as we'll see later in this chapter.

Sometimes, we can even feel like our brains are betraying us. For instance, it's frightening how easy it is to create false memories that feel completely real. Researchers have demonstrated the phenomenon over and over, but here's a simple example: In 2010, psychologists in Germany asked people to watch videos of other people performing simple tasks like shaking a bottle or unlocking a lock; two weeks later, many of the viewers remembered completing those actions themselves. It seems our brains can take empathy too far, logging in others' experiences as our own.

Even more disconcerting, many people can have the *same* false memory. In one example, nicknamed the Mandela Effect, thousands of people say they remember Nelson Mandela dying in prison in the 1980s, even though he actually lived until 2013. And I, along with lots of other people, remember a silly movie in the early 1990s called *Shazaam* that starred the comedian

---

* An average of 157 Americans a year died of constipation, according to U.S. CDC data from 2008 to 2017. Globally, sharks kill an average of six people a year, according to the International Shark Attack File.

Sinbad as a genie—a movie that never existed. I can even picture Sinbad on the movie poster, wearing the genie turban and standing with his arms crossed. Most likely, we all remembered another film in the same wrong way: the 1996 film *Kazaam,* which cast NBA star Shaquille O'Neal as a genie.

In the case of delusions, a false idea can also begin with something all too real—a maddening itch, say. Searching for an explanation, the itchy person might remember seeing fleas on the cat, and *poof,* the idea of an infestation is born. From there, the brain excels at finding the evidence it expects: a phenomenon known as confirmation bias. Our minds fill in gaps and make connections between experiences and memories; it's how we make sense of the world. But it also means that someone looking for evidence of an infestation will usually find it, and from there become convinced of something that doesn't exist.

But perhaps the biggest mental trick is the illusion that we're in control. Without getting into debates about free will (suffice it to say that some neuroscientists insist there's no such thing), we don't get to choose a lot that happens in our brains. Psychopaths don't decide to have abnormal connections between parts of their brains, and no one chooses to feel bugs that aren't there. Likewise, we can't control all of our abilities: As we'll see in one of the stories that follows, for instance, some of us are born with a highly tuned ability to recognize faces, whereas others can't pick lifelong friends out of a crowd.

If we're learning one thing about the brain and human psychology, it's that it's actually pretty normal to be a little abnormal in some way. Most of us have at least one irrational fear—or three (for me, it's crocodiles and getting hit by a bus). And we all have other mental quirks, too. One of mine, which scientists are just beginning to study, is a ramped-up aversion to particular sounds, called misophonia. For people with the condition, certain sounds like chewing or foot tapping trigger something akin to a fight-or-flight response, an intense feeling of anxiety and even anger. When I discovered that what I was feeling has a name and that

I'm not alone, I was finally able to start talking about it. Lifting the stigma around discussing our darkest thoughts is an important first step toward making sense of them.

And to be sure, even making sense of our own minds is a challenge. Take neuroscientist James Fallon, who we'll meet later in this chapter. He was studying the brains of psychopaths when he discovered that his own brain scan showed the same abnormalities. Looking back on his life, he realized he shared some of their personality traits as well. If a neuroscientist can fail to recognize himself as a psychopath, and doctors and biologists can convince themselves of insects that aren't there (yes, it has happened), maybe none of us should presume to truly know our own minds.

That can be a bit disconcerting. But then again, for the most part our weird brains are nothing to be afraid of; they're just another frontier to be explored. So perhaps these stories will spark something that resonates with you, or at least pique your interest—and in turn, keep us all exploring what's going on in our heads.

THE INVISIBUGS

How delusions work

The nightmare began in the summer of 2018: The chickens in the backyard coop had mites.

Lisa (not her real name) treated her chickens with a pesticide and saw some of the pepper-size mites land on herself. Weeks later, when she started feeling an itching, skin-crawling sensation, it made sense to think of the mites. From there, "it was a perfect storm," she says. The internet provided plenty of horror stories from people convinced that they—and everything they owned—were infested. She tried creams and lotions; nothing got rid of the feeling.

Lisa and her husband had planned a long hiking trip in Europe. But the feeling kept getting worse once they were overseas. She thought the mites might have gotten into her long hair, which she wore to her shoulders, so she chopped it short. The patches that delivered her hormone replacement therapy kept falling off, so she quit using them entirely. Misery set in. "I'm waking up in the middle of the night screaming at my husband that they're eating me alive," Lisa recalls. "I stopped sleeping, and I pretty much stopped eating." By the time the couple were loading their packs to head into the backcountry, Lisa's husband said, we can't do this. They returned home to Idaho. And that's when Lisa started what she calls "this thing with the house."

"I literally ripped the house apart," she says. Following protocols she had seen online to eradicate mites, she took down all her curtains. She stripped the bed, encased her mattress in plastic, and would wear only clothes that she had freshly washed and ironed; most of her clothes went into the trash. So did table runners, throw rugs, anything that was fuzzy, and some stuff that wasn't. She sprayed the house with essential oils and vinegar and vacuumed the floors at least three times a day. Still, nothing relieved her misery.

Lisa had already made the painful move of putting down her chickens, her beloved pets. She called an exterminator, who couldn't find any mites in the house. Her doctor said all her test results were normal. Yet still, her skin crawled. Finally, in desperation, Lisa moved out, leaving her husband alone in the house.

"At that point, I became suicidal," she tells me. She was isolated, afraid to go to work or to see her friends or family. She feared that she'd contaminate them with mites. Then her sister gave her the name of an expert who she thought might be able to help.

Gale Ridge, it turns out, was equipped to help, though not in the way Lisa expected. Ridge is an entomologist, an insect expert at the Connecticut Agricultural Experiment Station who specializes in the study of bedbugs. But she has also become one of the few intimately familiar with the condition known as delusions of infestation, the persistent belief that one's body is infested with small creatures when none can be found.

As entomologists will tell those experiencing such delusions, only two kinds of arthropods actually infest humans, meaning they live on us as parasites: lice and a particular species of mite that causes scabies, *Sarcoptes scabiei*. Both are fairly easy to diagnose when present. Lice are visible to the human eye; scabies mites aren't, but cause a characteristic pimply rash and can be seen by a doctor under a microscope. Then there are the teensy *Demodex* mites that live in our pores; some people might think of them as an infestation, but for the most part they're a normal occurrence on everyone's body, much like the bacteria that live in our guts. Other notorious "bugs," such as poultry mites, bedbugs, or fleas, don't live on humans; they may bite a person, but then they leave.

Ridge starts every case by looking vigilantly for actual insects or mites. If she doesn't find any, she must tread lightly. Many of the people who come to her are adamant that they're infested with something, and have already been frustrated by doctors who've turned them away; they generally aren't interested in being referred to a mental health professional. Over time, their lives start to revolve around their belief in the infestation, and proving it to others becomes paramount.

One case in point is a family that Ridge worked with in Massachusetts. A few years ago, Kelly (not her real name) had an uncle who began telling family members a horrifying tale: He believed he had bugs living inside him, ones with hard shells that crunched when he squashed them. He could feel them moving around, especially in his nose and private parts. At first, his family told him—gently—that it wasn't possible, but he just tried harder to convince them.

To collect "samples" as proof, Kelly's uncle would dig in his nose with tweezers, pulling out bits of tissue and cartilage until he had bored a hole through his septum; now he whistles when he breathes. After myriad tests turned up no sign of insects, his doctors seemed to give up. "All these medical doctors—not only were they not helping him, but they were embarrassing him, quite frankly," Kelly says. "He would walk out of there with his head down, so depressed."

One doctor prescribed him a medicine intended for people with delusional thoughts, but he refused to take it. "He said he doesn't need antipsychotic medicine," she says. "He needs samples to prove everyone wrong."

Finally, Kelly took her uncle to see Gale Ridge. Ridge examined his samples and met with the family in person to deliver exactly the news that Kelly's uncle didn't want to hear: The samples had no insects or living creatures of any kind. He left frustrated and angry.

For years, entomologists have insisted that these delusions aren't as rare as psychiatrists and the public may think. And in 2018, a study by the Mayo Clinic suggested they're right: The first population-based study of the condition's prevalence found that about 27 out of 100,000 Americans a year have delusions of infestation. That would mean around 89,000 people in the United States right now are plagued by the condition—and that's a conservative estimate, based on those people who sought help from a doctor.

The internet, meanwhile, has likely served to swell the ranks of those who are suffering. Blogs and websites maintained by people who believe they're infested give people who share this belief a sense of community—but also isolate believers from friends and family who don't go along with the idea. Many of these sites also spread misinformation and promote conspiracy theories that reinforce delusional thoughts, and often try to sell readers on sham solutions.

But as bizarre as the whole thing seems, it's actually not hard to see why people are so willing to believe that they're covered in insects or parasites they can't see. Often, an underlying condition really is causing the itching or crawling sensations that lead people to believe they're infested. Allergies, nutrition, stress, nerve conditions, and reactions to many common drugs can all be root causes.

But what often starts with an itch or other skin or nerve condition becomes a fixation, frequently on insects. That may be because most people are already a little afraid of insects, says Ridge. "So when people believe they've been bitten, they naturally default to that. It's almost instinctive."

Once they become convinced that they're infested, their stories tend to develop along similar lines, says Nancy Hinkle, the University of Georgia entomologist. One common thread among sufferers is a word they use: "desperate." "They call and say, Dr. Hinkle, you have to help me—I'm desperate," she says. Many entomologists have learned what's coming next and will say they can't help; a few, like Hinkle and Ridge, regularly take on such cases.

If an entomologist can't find any insects or parasites, and no other physical cause can be found for the person's symptoms, any doctor—even a family physician—can prescribe medications to treat delusional thoughts. But after six months or so,

Ridge says it tends to get harder to sway people toward such treatments. Another challenge is that the medications belong to a family of antipsychotic drugs, and the word "antipsychotic" is an enormous barrier for people suffering from this condition.

"A lot of patients won't take them," says dermatologist Mark Davis, an author of the 2018 Mayo Clinic study of delusional infestations. "They say, you're just saying I'm crazy, and I'm not." Prior to that research, Davis and his colleagues had reported on 147 cases of delusional infestation seen at the Mayo Clinic over seven years; he wasn't aware of anyone who had successfully overcome their delusions. Often, he said, people come to Mayo expecting to be diagnosed with an exotic new kind of infection and leave disappointed, never to be heard from again.

Another dermatologist, John Koo of the University of California, San Francisco, takes a different approach. Koo, who's also trained in psychiatry, doesn't try to talk patients out of the idea that they have an infestation—it doesn't work, he says. Instead, he offers them empathy. "You cannot agree with their idea, but you can certainly agree with their suffering," he told *Clinical Psychiatry News* in a video interview in 2018. Once patients are comfortable with their doctor, Koo says, they're more likely to try a low dose of an antipsychotic medication, such as pimozide, on a trial basis. And if they prefer to think it worked by killing the bugs rather than by acting on their brains, so be it.

As entomologists, Hinkle and Ridge are uncomfortable in playing along with the idea of nonexistent insects in any fashion. But both have plenty of empathy, which has taken an emotional toll. "Sometimes I can't turn it off," Hinkle says. "I can't go to sleep." She'll lie in bed remembering a woman she talked to that day, thinking, she's probably bathing in Clorox right now.

For Lisa, calling Gale Ridge helped her turn the corner. When the two talked on the phone, Ridge explained the biology of poultry mites and told her about all the other things that could be causing her itching, tingling skin. She reassured Lisa that the mites weren't infesting her body, but that she wasn't crazy; she just needed to get at the root cause of the itch. "For some reason," Lisa says, "I don't know what happened, but I started believing her. I think Dr. Ridge was part of saving my life, honestly."

The fact that Lisa had stopped taking her hormone medication around the same time she began to feel infested was a red flag. She returned to the one doctor she had trusted, an internist who was willing, as Ridge had recommended, to "think outside the box." They started working together intensively.

The doctor told Lisa that a swing in hormones like she experienced can cause formication, a skin-crawling sensation that would feel just like insects walking on her body. He got her back on the hormone patches she had stopped using, then met or spoke with her every day on the phone for a week, monitoring how she felt. Each day, he'd ask her to do one small thing to get back to normal: stop ironing her clothes or throw away the products she'd been using for her cleaning rituals. She started eating normal meals and going for walks.

Little by little, the idea of the infestation began to lose its hold. Finally, Lisa moved back into her home and was able to sleep in her own bed—though with a new mattress and bedding. She then went to a therapist and started cognitive behavioral therapy. When I talked to her about eight months after her ordeal had begun in the chicken coop, Lisa said she thought she was about 97 percent healed. She hasn't yet felt able to go out to the coop, and she's not sure if she'll ever have chickens again. But that skin-crawling feeling of formication did go away after she went back on the hormones, and she's relieved to have her life back. "It takes a village," she says—and Ridge was an important part of hers.

For Kelly and her uncle, Ridge was also a last resort—and part of a turning point. Once the family recognized what was really going on, they were able to stand firm the next time Kelly's uncle was in the hospital, and persuaded him to allow himself to be admitted for temporary psychiatric care. He received the medicines he needed and started to improve; Kelly reported to Ridge that once he went home, a nurse was assigned to visit daily and make sure he took his medications properly.

Only time will tell, Kelly says, if her uncle will ever be truly cured. But she thanked God, and Ridge, for giving him a chance. And she, like Lisa, hopes that telling her story might help others recognize what's happening to them or someone they care about—before the invisible bugs take over.

THE VOODOO DOLL RIDDLE

Why stabbing is oh, so satisfying

When Brad Bushman handed out voodoo dolls and told people to imagine their spouse, he expected that a lot of stabbing would ensue. As a social psychologist, Bushman knew that it's typical for people to act most aggressively toward the people they're closest to. But anyone who's been married probably could have predicted just as much.

Each member of the 107 couples who volunteered to participate in Bushman's 2014 study got a doll and 51 pins; each was instructed to stab the doll every night, depending on how angry they were with their actual spouse. (Supremely ticked-off husbands and wives used all 51 pins.) The couples participated in the experiment together and received $50 each for their services—but I bet that most of them would have done it just for the doll.

After all, something is strangely satisfying—even cathartic—about poking a voodoo doll. You get a little taste of revenge, but it's a lot safer than, say, poking your actual husband in the leg with a pin. And the practice works on more than just spouses. A 2018 study of office workers, in fact, found that stabbing a voodoo doll representing an abusive boss improved workplace morale. "Retaliating against the harm-doer," the researchers wrote in the *Leadership Quarterly,* "restores a sense of justice and thus affirms people's perceptions that those who do bad get what they deserve."

But as we've all experienced, just how much punishment we think our annoying boss or spouse deserves can vary. One day, you might be able to overlook your wife stealing the remote control—again—while another day it might send you through the roof. Bushman and his team wanted to understand the factors that make spouses lose their cool.

The team hypothesized that people might get more stabby with the dolls when they were hungry—perhaps because the brain requires energy to control behavior. And indeed, they found that when people in their experiment had low blood sugar, they went through a lot of pins. "Self-control of aggressive impulses requires energy, and much of this energy is provided by glucose derived from the food we eat," Bushman's team wrote in the *Proceedings of the National Academy of Sciences*. It's hard to say whether there's some particular breakdown in our brain's ability to control aggression when our blood sugar is low without further experiments—but the feeling is quite familiar. Many of us know it by another name: hangry.

Now it may seem obvious that being hungry makes us grumpier. But measuring how that translates into aggression—behavior intended to harm others either physically or psychologically—turns out to be more difficult than just asking people how aggressive they feel. Just as doctors have struggled to come up with a good way to measure a patient's pain, psychologists have long struggled to measure aggression. (You can't just measure violent acts because, thank goodness, we don't act on most of our aggressive impulses.) So how does one define how aggressive someone is, in a way that's repeatable and can be compared across groups of people, anywhere, anytime?

Psychologists love to come up with names for the various tasks they give people in experiments, and the test for aggression that Bushman used in the voodoo doll study has come to be known as the Voodoo Doll Task. It has turned out to be a great tool for studying aggression. People are given a doll that represents a specific person: either someone they know or someone who has provoked them in some way. The doll can be real or computerized—but either way, participants are told they can stab it with as many real or virtual pins as they want.

The Voodoo Doll Riddle

I tried the version that Bushman's team used, a virtual voodoo doll on a website called *Dumb.com*. I was skeptical at first: Could clicking on a picture of a doll on a screen really be satisfying? When I opened the web page, I was met with a realistic-looking picture of a doll made of white string, with orange string wrapped around it like a striped shirt. There's a line where you can type a name for your doll; I named mine after someone I felt wronged by (none of your beeswax). When the name popped up across the doll's chest, I admit I began to look forward to poking it.

Using my computer's trackpad, I picked up a virtual pin and hovered over the doll, relishing for a moment the power to choose where to poke. I clicked the doll's foot, and the pin pushed in. The doll's eyes watered for a moment, and its smile flipped to a frown. Whoa—I hadn't expected the doll to react. In addition to pins, the program also offers a lit candle and a pair of pliers. I touched the candle to the doll's arm, and it winced, wriggling a little as if in pain. Now its eyebrows curved in a look of worry, and I almost couldn't bring myself to try the pliers. But eventually I picked them up and clicked on the doll's belly. A small dark scar appeared, and the doll writhed in apparent pain for as long as I held down my click. At that point, it all seemed a little too real, and I felt not just ridiculous, but a little ashamed for hurting the doll. Then I felt silly for worrying about hurting a computerized doll—but of course, the whole point is that our minds so easily link the doll to the person it represents.

Before the Voodoo Doll Task, researchers had cooked up various other aggressive acts for people to commit in a laboratory setting. Among these was the Hot Sauce Paradigm. Participants armed with bottles of hot sauce are told to squirt as much as they want on a piece of food that, they are led to believe, another person who can't stand spicy food will have to eat in its entirety. The more hot sauce they add, the more aggressive they're deemed to feel.

That test worked pretty well, and researchers used it to study a variety of social psychology questions—for example, whether people who are particularly sensitive to rejection tend to become more aggressive when rejected (they do). But today, the Hot Sauce Paradigm, like so many before it, has more or less fallen by the wayside. For the ruse to work, study participants can't know one another well enough to realize that, in some cases, the subject of their hot sauce torture actually loves spicy food. Plus, these days a lot of psychology research is done using internet questionnaires—and even with cloud computing, there's no way to transmit hot sauce through the intertubes.

Ah, but voodoo dolls are another story. Not only can you furiously stab one in the privacy of your home, but as I also found, the practice can play out just as well on a computer screen. You also don't have to be anywhere near the subject of your aggression; in fact, it's better if you're not. In the study of hangry spouses, the subjects were directed not to do their stabbing in front of each other. One can only imagine the escalating retaliation the poor dolls might have endured otherwise.

For voodoo dolls to work as a test of our emotions, we have to treat them as though they are, in some sense, alive. And as it turns out, people tend to do exactly that. This brings us to another idea that goes back even further in psychology research: magical thinking. In 1986, Paul Rozin (who you may recall from earlier discussions of disgust) and his colleagues reported in an influential paper that modern Americans abide by the same two "laws of sympathetic magic" used in traditional cultures that had been documented in the early 1900s. Essentially, sympathetic magic is the idea that a person or thing can be affected by whatever happens to an object representing them. It's why burning someone in effigy makes a powerful emotional statement; in some sense, we feel like we're harming the person represented.

One law of sympathetic magic is contagion, the transfer of properties between objects. This is what makes celebrity-owned objects sell for gobs of money on eBay: The idea that

Kim Kardashian wore this bra! makes it more valuable. A negative quality can also be passed to an object; for instance, in some places food is considered contaminated if touched by a menstruating woman, who's considered unclean.

The other law of sympathetic magic is similarity: the idea that an image of an object equals the object. It makes poop-shaped chocolates a poor business venture, and it makes voodoo dolls satisfying even without any belief in the supernatural. The propensity of our brains to equate symbols to the real thing is how we connect a doll to the person it represents.

And in experiments, this kind of magical thinking does appear to play out in real life. In 2013, a team of psychologists and sociologists reported in the journal *Aggressive Behavior* that people will mentally transfer the characteristics of a person onto a voodoo doll representing that person. As a result, "the process of causing harm to a voodoo doll by stabbing it with pins has important psychological similarities to the process of causing actual harm to the person the voodoo doll represents."

So stabbing a doll is an excellent measure of aggression precisely because, somewhere deep in our lizard brains, we feel as though we are actually stabbing our husbands, wives, bosses, friends, neighbors, and so on.

Yet all this pin poking is ultimately for a good cause. As the proponents of the Voodoo Doll Task argue, these symbols can help researchers understand why and when people get aggressive—and hopefully guide us toward ways of reducing aggression. Scientists have found, for instance, that in addition to hunger, a lack of sleep can trigger aggression in married couples. And in any kind of social situation—they've tested many, from classroom to workplace to bedroom—the sting of rejection ramps up aggression and a desire to retaliate.

Even if you don't think of yourself as an "aggressive person," you've probably felt that emotion in all these circumstances. And in fact, psychologists say that aggression is one of the most misunderstood concepts in behavioral science. "It is commonly viewed by the general public as an aberrant form of behavior," wrote neuroscientist Robert Huber of Bowling Green State University and psychologist Patricia Brennan of Emory University in a 2011 book on the subject. And yet aggression is ubiquitous, even necessary to some extent—not only in humans, but also throughout the animal kingdom. A creature completely lacking aggression would be unable to defend itself or compete for resources.

The question for psychology and neuroscience is not whether we have a capacity for aggression, but how we keep it in check most of the time. Aggression run amok is counterproductive. But a finely tuned version that's unleashed to aid our survival has been key to our evolution as a social species.

So if a scientist ever hands you a voodoo doll, don't feel too bad about stabbing away. A little bit of aggression is, after all, only natural.

BACK OFF, BOZO

Turns out that clowns really are creepy

In the summer of 2016, clowns began creeping across America, lurking in the woods and freaking out the general public. In South Carolina, several children reported that a group of clowns had tried to lure them into the woods. In Florida, two clowns wreaked havoc by chasing people around wielding an ax and a baseball bat. Scary "clown sightings" became a meme, occupying a dark crevice in the popular imagination that UFOs once filled.

Soon, clown alerts were popping up in England, Australia, Canada, and Scotland, too. Whether prompted by teenage pranksters, a movie-marketing ploy, figments of the imagination, or real ones out to do harm, clowns are universally described with one word: creepy.

But what is creepiness, exactly? It's not quite the same thing as being scary or disgusting, though there's certainly an element of threat or potential harm. "A mugger who points a gun in your face and demands money is certainly threatening and terrifying. Yet most people would probably not use the word 'creepy' to describe this situation," says Frank McAndrew, a social psychologist who published the first large study on the subject in 2016. McAndrew explains that creepiness is a feeling of anxiety raised by an ambiguous situation—that is, when we're not sure if something is threatening or harmful.

And when we're not sure about a threat, our unconscious mind sends up a flare. Something seems off, maybe dangerous, so our brains quietly make associations behind the scenes. That's what we call intuition—and it's often a way of protecting ourselves.

So why would clowns be considered creepy? They're supposed to be silly, happy-go-lucky entertainers of small children, right? Scientists are giving this contradiction some thought—and clown sightings turn out to be great case studies on the nature of creepiness. "In many ways, clowns combine a perfect storm of freaky things," McAndrew observes.

Not only are they mischievous and strange looking, he continues, but you also can't tell who they are or what they're really feeling behind their painted-on smiles. This has led some people to speculate that clowns are creepy because they fall into the "uncanny valley," a not-quite-human appearance often ascribed to particularly realistic robots. The idea is that we like and feel empathy for robots that look somewhat humanlike (think C-3PO), but are repulsed by those that look too human. So, the thinking goes, clowns scare us because they blur the lines of looking like people—understandable, given the heavy makeup, huge feet, and bizarre hair.

But the uncanny valley seems at best a partial explanation for the creepiness of clowns. For one thing, the idea normally applies to objects that resemble humans: robots, dolls, ventriloquists' dummies, and those computer-generated characters that looked like Tom Hanks in the movie *The Polar Express*.* Clowns clearly *are* humans, though—weird-looking ones, but not facsimiles. And let's not forget that they haven't always been loathed. If the uncanny valley alone were the problem, we'd expect clowns and other heavily made-up actors to have always elicited feelings of creepiness. Yet there have been many beloved clowns, including Bozo and Ronald McDonald; I even remember when clowns were a perfectly acceptable decoration for a child's room. Try that now and see how your kids—much less, other adults—respond.

---

* The 2004 movie used a then new technology known as performance capture. A live actor, in this case Tom Hanks, wears a skintight suit studded with sensors that translate movements to a computer-generated image. The hyperrealistic result became a case study in the uncanny valley. CNN called the movie "at best disconcerting, and at worst, a wee bit horrifying."

According to McAndrew, what really makes clowns creepy is that they're ambiguous characters in so many ways. "If a person is willing to flout the conventions of society by dressing and acting as they do," McAndrew says, "what other rules might they be willing to break?"

Clowns have always been a type of trickster, according to author and expert Benjamin Radford. Radford argues that clowns have mixed horror and humor since the days of Punch and Judy, in which the harlequin-inspired Punch was a jokester who also beat his wife and killed his child. You never knew what he'd do next, which was part of his draw.

That tracks with McAndrew's research. "It is only when we are confronted with uncertainty about a threat that we get the chills," he observes. "It would be considered rude and strange to run away in the middle of a conversation with someone who is sending out a creepy vibe but is actually harmless. But at the same time, it could be perilous to ignore your intuition and engage with that individual if he is, in fact, a threat. The ambivalence leaves you frozen in place, wallowing in discomfort."

Clowns, as pranksters, have always toed a line between entertainment and mischief. So it's easy to flip the traditional, happy trope on its head, as we often do in literature and film. In 1892, a murderous clown made his debut as star of the opera *Pagliacci*. And the evil clown trope took off in the 1970s and '80s after serial killer John Wayne Gacy entertained at events as Pogo the Clown, and Stephen King's *It* became a global best seller. Today, clowns seem to be portrayed more often as scary than as funny.

And that brings us back to the onslaught of creepy-clown sightings. First, people started posting photos and videos of real sightings on YouTube and social media; then the phenomena evolved and spread as a series of online memes, including pictures of clowns accompanied by scary mottoes like "Sleep tight, I'm under your bed." The hashtag #IfISeeAClown, where people joked about how they'd react if a clown appeared, prompted hundreds of tweets. *It* author Stephen King even stepped in on Twitter to defend clowns at the height of the paranoia. "Most of 'em are good, cheer up the kiddies, make people laugh," he tweeted.

Harvard computer scientist Michele Coscia, who has studied how ideas go viral, thinks he might know why creepy clowns were such a hit on social media. He looks at a value called "canonicity," a measure of how unusual a meme is. Lower canonicity means the idea is more unusual, and unusual ideas are the most likely to go viral. In the case of creepy-clown sightings, Coscia notes that there were already prank memes and memes featuring creepy clowns—but not the conjunction of the two: pranks where people dressed up as creepy clowns. And *boom:* A new, low-canonicity idea was born, primed to go viral.

Then social psychology kicked in. "Social media fans the flames by giving us a false sense of how widespread something is, and how threatened we should be feeling," McAndrew says. "Better to err on the side of caution by protecting your children from killer clowns than to err in the other direction. We now have the ability to sound alarms and spread rumors with a megaphone, and we never pass up the opportunity to do so."

Based on McAndrew's research, the resurgence of creepy clowns seems inevitable. And clowns aren't likely to lose their spookiness anytime soon, either. In the course of his research, Coscia asked people to rate the creepiness of different professions. The winner, edging out taxidermists, sex shop owners, and funeral directors? Clowns.

NEVER FORGET A FACE

Inside a special crime-fighting superpower

On August 28, 2014, a 14-year-old girl named Alice Gross stepped out the door of her West London home for a walk and never returned. The search to find her became London's largest since its 2005 hunt for terrorist bombing suspects. Neighbors tied thousands of yellow ribbons around every tree, post, and railing in Gross's Ealing neighborhood, and hundreds of police officers looked for her. Their only clue: grainy closed-circuit footage that showed the girl walking along a canal a few hours after leaving home. Fifteen minutes after she passed the camera, it recorded a man riding a bicycle in the same direction.

London is peppered with an estimated half a million CCTV cameras, mostly privately owned, that monitor homes, businesses, and public places. They've become a prime resource for police—but the footage is useful only if someone can recognize the suspect. In the Alice Gross case, officers followed the man on the bicycle as he appeared on a string of cameras after the girl's last appearance at the canal. He disappeared for a time near a wooded area, then reemerged and bought beer in a shop, then vanished again. It seemed suspicious. Who was he?

That's where law enforcement had an advantage. In 2015, London's Metropolitan Police Service officially formed a team of super-recognizers: men and women with a preternatural ability to distinguish and match faces. When Gross was kidnapped in 2014, the police were already experimenting with this group; with their help, officers had identified the faces of criminals who'd had previous encounters with law enforcement, such as Stephen Prince, a bomb-throwing rioter who was arrested in 2011.

After studying the footage of the man trailing Alice Gross, the budding super-recognizer task force identified him as Arnis Zalkalns, a 41-year-old builder who had served seven years in a Latvian prison for killing his wife and had been questioned in the 2009 sexual assault of a 14-year-old girl. Police launched a search for Zalkalns and found him dead in the woods, where he had hanged himself; Gross's body had been found four days earlier, a few miles away in a river. After further investigation, including DNA evidence, police and Scotland Yard were confident that Zalkalns had killed Gross, dumped her body in the river, and then killed himself. Tragically, though it was too late to save her, the super-recognizers had played a key role in solving Gross's murder.

Detective Chief Inspector Mick Neville is in awe of the group's abilities. "Gary Collins is so good that he ID'd three people over his Sunday roast," he said of one officer who would relax on weekends with an iPad loaded with photos of criminal suspects. After London's 2011 riots, the unit combed through thousands of hours of footage; Collins alone identified a staggering 190 rioters.

These face-spotting stars normally work in their local stations, building up a mental library of the area's criminals, and periodically join New Scotland Yard to help solve crimes. So far, their big wins include a serial molester on different city bus routes (he was tracked to particular routes and arrested), a jewel thief, and the Russian spies accused of poisoning former British double agent Sergei Skripal and his daughter in 2018.

Every year, London's super-recognizers are out in force at big events like the Notting Hill Carnival, Europe's biggest street fair. Crime has been a problem, so during the festival the unit sits in CCTV control centers and scans the crowd, looking for known troublemakers. In 2015, they spotted members of rival gangs edging close to one another. Officers on the scene located and disarmed the men, averting a potential fight. "The senior detective was amazed at their ability to spot suspects in dense crowds," Neville says.

And now, there's a private market for these skills. Neville went on to head up London's Central Forensic Image Team, which has tested thousands of police officers and identified the super-recognizers among them. He then became CEO of Super Recognisers International, a private investigation firm that specializes in finding missing people and verifying identity. Clients include Britain's football clubs, for which the company identifies "banned supporters, ticket touts, and known trouble-makers," according to the company's website.

The ability to recognize faces falls along a spectrum, according to David White of the University of New South Wales's forensic psychology laboratory. At the lowest end are people who have "face blindness," a condition known as prosopagnosia. (Oliver Sacks, the celebrated neurologist who died in 2015, suffered from this; he wrote in the *New Yorker* that he recognized his close friend Eric by his "heavy eyebrows and thick spectacles.") On the other end are super-recognizers, who in some cases can identify someone after seeing them just once, even years later.

"These 'super-recognizers' are about as good at face recognition and perception as developmental prosopagnosics are bad," a team of scientists from Harvard and University College London revealed in one of the first studies of the phenomenon, published in 2009. The group examined four super-recognizers and compared their facial recognition abilities to those of 25 control subjects. The super-recognizers outperformed the "regular" recognizers in every test the scientists gave them—in some cases with perfect scores.

Most people overestimate these skills, White says. "When people think of face recognition, they think [it applies to] recognizing people they know," he says. But recognizing the face of a stranger, based on two different photos laid side by side—well, that's much, much harder.

If you'd like to get an idea of where you fall on the spectrum, Josh Davis, a psychologist at the University of Greenwich, devised a short, simple quiz, as well as a more detailed test that will help researchers map out just how many people fall along each part of the spectrum.[*]

So far, Davis says, it appears that face recognition is an innate ability—not learned, for the most part—that's distributed on a bell curve, like IQ. No particular genes have been linked to super-recognizers' ability, but a 2010 study found that face-recognition ability was extremely similar in identical twins, compared with fraternal twins: a first indicator of a genetic link. Similarly, face blindness has what's known as a developmental form, which is present from birth and can run in families.

As to how it works, prosopagnosia is tied to the fusiform face region: slivers that run along the bottom of the brain near the back of your head. That's probably a good place to look for evidence of unusual brain activity in super-recognizers as well.

One of the most pressing questions for scientists has been whether there really is something special about the way our brains process faces, as opposed to other objects. Oliver Sacks struggled not only with faces, for instance, but also with buildings; he wrote that he once got lost and wandered past his own home several times, not recognizing it.

Yet this scenario is not true of all prosopagnosics, and a number of experiments suggest that our brains do handle faces as a special category—perhaps because they're so important to us as social creatures. In 2017, researchers at Penn State University used functional MRI to measure brain activity while people identified faces and objects. They found that face recognition and object recognition utilized different parts of the brain, and

....

[*] You can find the simple test for super-recognizer skills here (available in English, Spanish, French, German, and Portuguese): www.superrecognisers.com.

that people with superior face-recognition skills used more of their brains when recognizing faces—not just the fusiform face region, but other parts as well.

Likewise, studies have found that although our brains first process features such as the eyes, nose, and mouth individually, recognizing a face is a holistic process that involves analyzing individual features, putting them all together, and examining the relationships among the parts. Some super-recognizers seem to excel at these tasks and are able to easily find a face in a lineup, whereas others seem to have a particular knack for remembering faces as well.

So how common might these super-brained super-recognizers be among us? Davis says that depends where you decide to draw the line. But tests so far suggest that about 2 percent of people have prosopagnosia—and about the same percentage are exceedingly good at recognizing faces.

How to best take advantage of people with natural face-sighting superpowers? The London police have garnered interest from law enforcement officers around the world, and are sharing their newfound knowledge with their colleagues. The University of New South Wales' David White has worked with the Australian passport office to develop training for its officers (whose job is to determine whether a stranger in front of them is the same person shown in a tiny passport photo).

White's research group tested a group of crack forensic examiners who specialize in face image analysis to determine whether their training has helped them identify faces more accurately than untrained people; if so, that training might help to improve the skills of people who match faces for a living. The experts, it turned out, did perform better than either untrained students or forensic experts who don't match faces regularly. (And by the way, they also outperformed facial-recognition computer algorithms on certain tasks; computers do well when faces are lit well and positioned the same way, like in passport photos, but often fail in less than optimal conditions.)

It's possible that people with above-average skills gravitated into these positions based on their abilities. But surprisingly, the expert face matchers showed another ability, unusual in those with "natural" face-matching skills: identifying faces that were shown upside down. To White, that suggests that the experts' training, which emphasized breaking down a face into component parts and matching each one, may be boosting their skills beyond their natural abilities.

A bit murkier, though, is how super-recognizers' identifications will hold up in court. In the United States, judges use a legal precedent called the Daubert standard to determine whether evidence is scientifically supported; as a result, eyewitness IDs of all sorts have come under scrutiny. But Neville says his team's matches are usually just the starting point for developing a criminal case, pointing police toward a suspect who can then be investigated using DNA or good old gumshoe police work. In these cases, the ID is more of an investigative tool than direct evidence of criminal misdeeds.

Regardless of the future of super-recognizers in law enforcement, learning of their existence has opened a window into how our brains work. Identifying other people is a fundamental skill for a highly social species like our own—and at the very least, understanding where we fall on the spectrum could help us defuse some awkward social situations. I know a woman who informs colleagues that she has face blindness, for instance, so they won't be offended when she doesn't recognize them at meetings. On the flip side, one of the super-recognizers in the 2009 study had this to say: "I do have to pretend that I don't remember [people], because it seems like I stalk them, or that they mean more to me than they do when I recall that we saw each other once walking on campus four years ago in front of the quad!"

The research is also revelatory for those who, like me, score a paltry 7 out of 14 on the simple face-matching test, which is in fact pretty normal. Though I'm far from having face blindness, I struggle to recognize actors in movies, and I often fail to recognize acquaintances if I see them out of context. (I'm terrible about walking right past co-workers on the street, for example, because I recognize them only in the office.) My husband, on the other hand, amazed me recently by spotting our cat-sitter at a coffee shop—never in a million years would I have recognized her. I had such faith in his abilities that I was shocked when he scored a 7, too. Now that I know we may not be so far apart, I suppose I'll have to actually pay attention at parties instead of just relying on him to remember people. So if I don't recognize you, I apologize: I'm only a 7.

CELLULOID PSYCHOS

Which film monsters are most realistic?

Fun fact: Real-life violent psychopaths aren't usually wild-eyed serial killers who laugh maniacally as they murder people, as they do in campy films. But the character of Anton Chigurh from the Coen brothers' *No Country for Old Men* is a little closer to the mark. He quietly walks up holding a bolt gun, and by the time you can say, "Hey, what's that thi—" . . . it's *ka-chunk,* and you're dead.

That's more what an actual psychopath—a person who feels no empathy for others—is like, according to forensic psychiatrist Samuel Leistedt, who has interviewed and diagnosed many of them. Although in pop culture we tend to equate psychopaths with crazed killers, Leistedt describes them as cold-blooded. "They don't know what an emotion is," he observes.

In 2014, Leistedt and his colleague Paul Linkowski reported on the three years they spent watching 400 movies looking for realistic portrayals of psychopaths. (This means that in the name of science, they not only viewed Alfred Hitchcock's legendary *Psycho,* but also sat through *Pootie Tang.*) Learning to diagnose a psychopath isn't easy, Leistedt says. For one thing, students get limited chances to interview real-life cases as part of their training—and many start out with misconceptions about what a psychopath is.

Based on their portrayal in popular culture, most people believe psychopaths to be dangerous criminals, even serial killers. But in reality, most violent offenders aren't psychopaths, and most people who exhibit psychopathic traits aren't criminals. Psychiatrists define psychopaths not by violent behavior, but by a suite of extreme personality traits. They're superficially charming, deceptive, manipulative, impulsive, and grandiose, meaning they have an outsize sense of superiority. They lack empathy for others and don't feel guilt or remorse. The condition is usually diagnosed using the Hare Psychopathy Checklist-Revised, or PCL-R, devised by criminal psychologist Robert Hare. A score of more than 30 out of 40 is considered diagnostic of psychopathy, which Hare has estimated to occur in perhaps one percent of the population.

And yet, psychopathy is not listed as a mental illness in the *Diagnostic and Statistical Manual of Mental Disorders* (or DSM), the authoritative guide for American psychiatry. Instead, the manual describes antisocial personality disorder, which shares similarities with psychopathy but is defined less by personality traits and more by a pattern of behavior that includes illegal acts. (Perhaps as many as 50 to 80 percent of prisoners meet the criteria for antisocial personality disorder, but only about 15 percent qualify as psychopaths based on the PCL-R.)

As for the causes of psychopathy, there's long been a debate about biology versus environment, or "nature versus nurture." Certainly both are important, and some researchers believe that people with psychopathy are either born (essentially "hardwired" in the brain) or made—often as children raised in violent, chaotic environments.

However psychopathy is created, neuroscientists have learned a bit about what sets people with psychopathy apart. For instance, scientists have found parts of the brain that appear to function differently in people with psychopathy than in the average person—including the amygdala, which is important for processing emotions. Some studies have also documented that the parts of the brain linked to empathy and self-control are less active in those with psychopathy. Moreover, their brains' reward system seems to be on overdrive, making them thrill seekers as well. Still, much remains unknown; it's not clear, for instance, what role genes play in psychopathy. So far, no particular gene or suite of genes has emerged that "creates" psychopathy.

Bearing in mind all that's known on the subject so far, Leistedt and his team decided to "diagnose" fictional psychopaths in movies in an effort to establish accurate portrayals and to understand popular misconceptions. Using the PCL-R and classifications outlined by forensic psychologist Hugues Hervé and psychiatrist Benjamin Karpman, Leistedt and other psychiatrists started reviewing films. First they weeded out clearly unrealistic characters, such as those with magical powers or who were invincible; that whittled down the original 400 films to 126 made between 1915 and 2010, portraying 105 male and 21 female* potential psychopaths.

Ultimately, the films served as a kind of social history of how psychopathy has been viewed since the early 20th century. In the early days of Hollywood, psychopathy was generally portrayed with caricatures of villains: gangsters, murderers, and mad scientists with bizarre affectations and a compulsion to kill for no apparent reason.

Later, real-life cases of serial killers came to light, starting with Ed Gein in 1957, and moving on to Ted Bundy and Jeffrey Dahmer. As this kind of criminal psychopathy came to be better known, movie psychopaths became more like them—or at least, like the parts of their behavior that Hollywood directors found compelling. We got the sexually deviant misfit, like Norman Bates in *Psycho* (1960) and the chaotic mass killer of the slasher genre, like Michael Myers of the *Halloween* series.

Finally, as Leistedt and his team reported in the *Journal of Forensic Sciences* in 2014, a growing fascination with the minds

----

* Psychopathy is generally believed to be more common in men than in women. However, in recent years researchers have pointed out that the behavior traits used to diagnose psychopathy may be more characteristic of male, as opposed to female, psychopaths. As a result, there's no firm breakdown.

of psychopathy has led to more nuanced, and sometimes realistic, portrayals of their behaviors and personalities. Instead of giggling killers with facial tics, at least a few of today's depictions have more depth, giving a "compelling glimpse into the complex human psyche," they observed.

In recent years, Hollywood has been perhaps most enamored of the so-called "successful psychopaths": those who use their psychopathic traits to their own advantage, often gaining the trust of others and then using them to acquire money or power. Leistedt notes that the fascination took hold in the wake of the financial crisis and high-profile trials like that of Bernard Madoff. (Apparently, vicious stockbrokers are the new bogeymen. Instead of disemboweling their victims, they gut their bank accounts.)

Here are a few examples of the best and worst film psychopathy portrayals, according to Leistedt's study.

FRIGHTENINGLY REALISTIC:

Michael Corleone, *The Godfather: Part II* (1974)
The Mafioso is a classic example of the criminal psychopath, which Leistedt and his colleagues say is probably one of the most common types of psychopathy to appear in films. As a young man, Corleone doesn't want to enter the family business, but as a Mafia boss, he becomes a ruthless killer.
Diagnosis: Secondary, macho psychopathy

Gordon Gekko, *Wall Street* (1987)
Gekko is a realistic portrayal of a successful psychopath, Leistedt says "probably one of the most interesting, manipulative, psychopathic fictional characters to date," he and his colleagues conclude in their paper. Leistedt adds that the story of real-life con man Jordan Belfort in the 2013 film *The Wolf of Wall Street* makes for an interesting portrayal as well: "These guys are greedy, manipulative, [and] they lie, but they're not physically aggressive."
Diagnosis: Primary, manipulative psychopathy

Anton Chigurh, *No Country for Old Men* (2007)

This contract killer hauls around a bolt pistol attached to a tank of compressed air: a handy tool for shooting out locks in doors and for shooting people in the head. Leistedt says Chigurh is his favorite portrayal of a psychopath. "He does his job and he can sleep without any problems. In my practice I have met a few people like this," he admits. In particular, Chigurh reminds him of two real-life professional hit men he interviewed. "They were like this: cold, smart, no guilt, no anxiety, no depression."

Diagnosis: Primary, classic/idiopathic psychopathy

SCARY, BUT NOT REALISTIC:

Norman Bates, *Psycho* (1960)

After the 1957 arrest of real-life serial killer Ed Gein—a case involving cannibalism, necrophilia, and a troubled maternal relationship—horror films about serial murder took off. The character of Norman Bates was inspired in part by Gein, launching a genre dedicated to portraying misfits, often with a sexual motivation to kill. This kind of behavior became closely linked to psychopathy, but Gein was more likely psychotic, meaning out of touch with reality. Psychosis, which is a completely different diagnosis from psychopathy, often involves delusions and hallucinations.

Diagnosis: Pseudopsychopathy, psychosis

Hannibal Lecter, *The Silence of the Lambs* (1991)

Yes, he scares the bejesus out of me, too. But Lecter's almost superhuman intelligence and cunning are just not typical among, well, anyone, let alone psychopaths. Lecter is a perfect example of the "elite psychopath" that became a pop-culture icon in the 1980s and 1990s. This calm, in-control character has sophisticated tastes and manners (think Chianti and jazz), an exceptional skill in killing, and a vain and "almost catlike demeanor," the researchers write, adding, "These traits, especially in combination, are generally not present in real psychopaths." In fact, the sense of grandiosity in people with psychopathy doesn't square with studies of their actual IQs, which show no clear difference in their intelligence compared with the rest of the population.

Diagnosis: Lecter's unrealistic blend of skills and traits makes him an ideal villain, but undiagnosable.

Movies aside, successful, nonviolent psychopathy can be found in all walks of life, as prominent neuroscientist James Fallon would discover the hard way. Fallon was studying the brains of people with psychopathy when one of the PET scans he'd done for a different study, on Alzheimer's disease, jumped out at him. It displayed the exact features he'd discerned in psychopaths—interesting, until he realized he was looking at his own brain. (Fallon had served as one of the controls for his study, with no idea he would uncover something so disturbing.)

When he considered his past behavior, Fallon realized he does possess some psychopathic[*] traits, including risktaking. But he's also had a successful career and stable relationships, and he's certainly not a killer. Instead, he considers himself a "pro-social psychopath," a term that some psychologists have used to describe people who have trouble feeling empathy for others, but are able to behave in socially appropriate ways.

Certainly, not all psychopathy is created equal, and current research backs the idea that people with the condition can lead very different lives, depending on their circumstances and particular blend of traits. Thus Fallon, who had a good childhood and achieved a high level of education, became a neuroscientist instead of a criminal. Likewise, people who score high on a scale of fearless dominance—a suite of traits found in psychopathy—can often be successful leaders in the workplace. But those who

[*] Fallon has also noted that he doesn't have "full-blown" psychopathy, meaning he scored below the commonly accepted cutoff for psychopathy of 30 on the PCL-R.

fall into a category that psychologists call "self-centered impulsivity" tend to act out and get themselves into trouble.

Scientists had long considered these troublesome psychopaths a lost cause; because their brains are not equipped to feel empathy, the feeling went, they'd never grow a conscience. But there have been glimmers of hope for at least some children who show signs of developing psychopathy. Punishing these kids has no effect on their behavior, so at the Mendota Juvenile Treatment Center in Wisconsin, researchers are taking a different approach. Staff at the center compensate kids for good behavior, no matter how small, taking advantage of the psychopathic brain's amped-up reward system. The idea is to teach them to behave in socially appropriate ways, even if they don't actually develop empathy.

It may be helping. Of 248 juvenile delinquents tracked for nearly five years after their release from correctional centers, the 101 treated at Mendota were around half as likely to commit violent crimes. Most important, none of them were arrested for killing anyone during that time, although kids from other facilities killed 16 people.

Not killing people may sound like a low bar—but it's an important one. As we learn more about psychopathy and how the brains of these people work, the hope is that we'll develop new treatments. At the very least, we might learn to depict them more accurately in pop culture, which could help us spot the ones we may encounter in real life. After all, there's no need for a maniacal laugh. Someone with true psychopathy, who feels no empathy and no remorse, is scary enough.

TYPES OF PSYCHOPATHY

Primary versus secondary psychopathy: Primary psychopaths are deficient in affect, or emotion, from birth, suggesting a genetic basis. They are often described as more aggressive and impulsive. Those with secondary psychopathy have been shaped by their environment, may have had an abusive childhood, and are often described as having more fear and anxiety than primary psychopaths.

SUBTYPES:

Classic/idiopathic: Score the highest on all sections of the widely used Hare Psychopathy Checklist-Revised, or PCL-R, showing low fear, lack of inhibition, and lack of empathy

Manipulative: Tend to be good "talkers" and associated with crimes involving fraud

Macho: Lack the glibness and charm of the previous groups but manipulate through force and intimidation

Pseudopsychopaths: Also called sociopaths; show antisocial behavior but score lowest among these groups on the PCL-R

PSYCHOPATHY CHECKLIST

The most commonly used checklist for diagnosing psychopathy is the Hare Psychopathy Checklist-Revised. It assesses 20 traits, and subjects are scored for each, using zero (does not have the trait), one (somewhat applies), or two (definitely applies). These are the 20 traits of psychopathy:

- glib and superficial charm
- grandiose (exaggeratedly high) estimation of self
- need for stimulation
- pathological lying
- cunning and manipulativeness
- lack of remorse or guilt
- shallow affect (superficial emotional responsiveness)
- callousness and lack of empathy
- parasitic lifestyle
- poor behavioral controls
- sexual promiscuity

- early behavior problems
- lack of realistic long-term goals
- impulsivity
- irresponsibility
- failure to accept responsibility for own actions
- many short-term marital relationships
- juvenile delinquency
- parole violations
- criminal versatility

THE SOUND AND THE FURY

Why certain noises annoy

t is hard to describe how I feel when I hear someone chewing
gum, but here's an analogy: It's as if someone has magically
reached through my skull and is inserting tiny needles, one by
one, into my brain. With each smack of the lips, another needle
jabs* me. *Smack, smack, smack.* It fills me with anxiety, a
gut-dropping sense of dread, and irritates me to the point of
anger. It's fight or flight, and I feel an overwhelming urge to do
both at the same time.[†]

I know it's ridiculous—it's just gum!—but certain sounds such
as loud chewing and knuckle cracking make me feel like an animal
trapped in a cage. I suffered for years before learning that what
I was feeling has a name: misophonia, a little-known condition in
which certain "trigger" sounds induce anger and anxiety. This
isn't the same as simply finding certain sounds annoying, which
we've all experienced. Instead, for a person with misophonia,
particular sounds—sounds that other people may or may not find
annoying at all—set off a specific and intense emotional response.

Lucky for me, I've been able to use coping mechanisms like
background music and earplugs to get by. But some people with
more severe misophonia can't share a meal with friends and
family, or lose relationships because of a sensitivity to their
partner's inevitable noises. And like me, many people who suffer

---

* This isn't a literal, physical sensation, but more about how it makes me feel—like the
person making the sound is doing something awful, and they're doing it to me.
† To be clear, I've never felt violent—just an urge to shout something rude about
someone's thoughtless and disgusting mouth noises.

its effects haven't been diagnosed formally. Misophonia mostly flies under the medical radar. It's not officially recognized as an illness; there's no universally accepted set of criteria for diagnosing it. And most doctors have never even heard of it.

Yet over the last decade or so, misophonia has started to gain recognition among psychologists, audiologists, and neuroscientists as a "real" condition. Support groups have popped up, therapists have begun experimenting with treatments, and brain research is revealing the neurological roots for what some have called "sound rage" (though many prefer the milder "selective sound sensitivity syndrome"). Some of the most important neurological findings to date have stemmed from questions by people with the condition who have reached out to scientists asking for help.

Miren Edelstein first came across misophonia in 2011, when one of her graduate advisers, a well-known neuroscientist named V. S. Ramachandran, started receiving queries from sufferers. Ramachandran studies how the brain handles sensory inputs, and is famous for his work on phantom limb pain. He had helped ease amputees' suffering by inventing a simple mirror box that tricks the brain into thinking a missing limb has returned. Misophonia sufferers hoped he might be able to figure out if their brains were similarly miswired.

Edelstein was studying auditory perception and cognition, and in particular perfect pitch—a far cry from misophonia, but related enough for Ramachandran to ask her to look into the mysterious new condition. The first thing she noticed was that although misophonia literally means a "hatred of sound," people with the condition don't hate sound or loud noises in general—only certain sounds, and only ones they've heard repeatedly over time. "That was curious," Edelstein says. "Basically everyone seems to have their own unique set of triggers."

She concluded that the angry, anxious response to trigger sounds seemed a lot like a fight-or-flight response: the physiological reaction that prepares us to deal with danger. Whether you're confronted by a 500-pound gorilla or are facing a final exam, your body reacts in much the same way. Your sympathetic nervous system kicks in, triggering your adrenal glands to produce hormones that cause your heart to race, your breathing to speed up, and your skin to sweat. Psychologists have found that to some degree, they can track our emotional state by monitoring the sweat from our hands, because the sweat response is most sensitive to our thoughts and feelings. (They don't even have to measure the actual sweat; the amount of electricity conducted* across our skin just slightly increases when our sweat glands produce moisture.)

Edelstein started measuring this skin conductance response in people with and without misophonia, to determine whether those with the condition were having an outsize physiological response to certain sounds. She played dozens of recordings of everything from birds singing and children laughing to whale songs and fingernails on a chalkboard. (Sounds were played both alone and in videos to separate the auditory and visual effect.) Pretty much everyone reacted physiologically to the sound of nails on a chalkboard or a baby crying. But sure enough, people with misophonia had strong reactions to other sounds like gum chewing, pen clicking, and the crunch of an apple being eaten.

The results were the first experimental proof that people with misophonia were experiencing sounds differently from other people. Not only did subjects with misophonia consistently rate certain noises more aversely than control subjects did, but their skin conductance measurements showed that they were reacting physiologically to those noises. The response was visceral and

....

* Skin conductance is one of the factors measured by polygraphs, or lie detectors, along with heart rate and respiration. The American Psychological Association has maintained for years that polygraphs are unreliable at detecting deception. However, this doesn't mean that skin conductance is useless; it has been proven to reliably show changes in a person's state of physiological arousal—just not whether they're telling the truth.

autonomic, meaning that it happened beyond the subjects' conscious control. This was an important finding for people with misophonia, who are often frustrated by well-meaning suggestions like "just ignore it" or "don't let it bother you." For people suffering from this condition, that's not an option.

In the course of her experiments, Edelstein noticed another quirk that many people with misophonia have reported: a stronger response to sounds made by people they're close to, like family or friends, as compared with strangers. That's unfortunate for people with misophonia and the people they live with, but Edelstein also thought it was intriguing: Somehow the context seems to be important to how a sound is perceived.

"Early on," Edelstein says, "a woman came in with misophonia and we played her audio of a baby eating this messy, really wet baby food. She couldn't see the video, just heard the sound. She was like, 'that's just awful,' and she was feeling very triggered." Then the researchers showed her the video version, and she could see that a baby was making the sounds. "Her response completely changed. Suddenly it was not nearly as bad, and she even said it was kind of cute."

With that in mind, Edelstein is now looking more closely at the time line of people's physiological response to trigger sounds to see if a "good" context, like the baby eating, has no effect—or if the subjects are still triggered but quickly able to override the signal.

Although this documentation is useful, it's still not clear what's happening in the brains of people with misophonia as they hear a trigger sound. Once Edelstein and her colleagues established a physical basis for the condition, the next step was to look for a biological mechanism: how hearing a sound triggers an emotion. And one research team has been using brain imaging to do just that.

Tim Griffiths, a neuroscientist at Newcastle University, began studying misophonia in 2012, after releasing a study that measured people's reactions to aversive sounds.* Like Edelstein's adviser, Griffiths began receiving messages from people with the condition who were looking for help.

At first, Griffiths thought the symptoms sounded like non-sense; he assumed these people might be neurotic, but not neurologically different from anyone else. Still, the idea was intriguing, so he started inviting some sufferers into his lab. "It was immediately clear that I was completely wrong," he says.

The people with misophonia described a uniform set of responses to certain sounds—but the sounds themselves were all over the map. For many, mouth noises such as chewing or sucking were the main problem, while for others, clicking noises sent them through the roof. One high-flying financial executive said he couldn't work in an office with an open floor plan, because of all the typing. Clearly, this wasn't the same phenomenon Griffiths had seen with the generally aversive sounds. The sufferers' hearing was normal, they had no other neurological condition, and their brain scans looked normal.

Perhaps, Griffiths thought, something was going awry with the processing of sound in the subjects' brains—something that wouldn't show up on images of the brains' structures, but might become clear using functional MRI. This technique uses blood flow to give an indirect measure of the activity in different parts of the brain. Griffiths started looking at the activity in brains of people with misophonia when they heard their trigger sounds, and compared that data to activity in people without misophonia. He also observed the reactions of both groups to the generally aversive sounds he'd studied previously.

....

* In that study, the most unpleasant sounds occurred in the frequency range of 2,000 to 5,000 hertz—a very sensitive range. Out of 74 sounds, nails on a chalkboard rated as the fifth worst, just above a female scream. The worst was produced by dragging a knife across a bottle. The more unpleasant the sound was, the more activity there was between two parts of the brain: the auditory cortex, which processes noise, and the amygdala, a part of the brain involved in fear and other emotions.

In the brains of people with misophonia, Griffiths found a pattern when trigger sounds were heard: increased activity in a cluster of gray matter, called the anterior insular cortex, that's tucked into a large fold on each side of the brain. This wasn't the same area that reacted strongly to noises that everyone hates— the aversive ones—but a different phenomenon altogether.

As Griffiths explains it, we're constantly bombarded with sounds, and our brains are constantly determining which are important (for instance, we've learned to ignore the hum of the refrigerator but pay attention when a door slams). The brain makes these decisions using a network of structures that communicate with one another; neuroscientists call it a "salience network." The anterior insular cortex is a crucial component, and that's the part of the brain that was overactive in people with misophonia.

So in the brain of a person with misophonia, it appears that a signal normally weeded out is deemed important, and is moved along accordingly. And tellingly, one of the abnormal links goes to the medial frontal lobe, which is involved in controlling our emotional reactions. "What we think is going on is an abnormality in this control system," Griffiths says. In the part of the brain that should be filtering out unimportant sounds, the accelerator is pressed, while in the part of the brain that should be tamping down an emotional response, the brakes are off. In the end, an otherwise innocuous sound ends up seeming like a threat.

Understanding this unusual condition can also help us understand more broadly how the brain links senses and emotions. "It's a natural experiment in emotional control," Griffiths adds. So far, his team's research supports the idea that control mechanisms in the frontal lobe normally suppress emotional responses to stimuli that don't warrant a response.

More research lies ahead to translate our understanding of misophonia into a plausible treatment approach. Down the line, it may be possible to "retrain" the brain's response to a sound—but just how to do that isn't yet clear. Some people report success with cognitive behavioral therapy, a type of psychological training that aims to change dysfunctional thoughts, emotions, and behaviors. Others say they've been helped by apps and other means of exposure therapy, which aim to gradually build up a tolerance to trigger sounds. But because none of the current methods has been rigorously tested in an experimental setting, it will be hard to determine what works, and for whom, until more work is done.

Last summer, Edelstein attended a retreat for misophonia researchers put on by the REAM Foundation, which has also put out a call for grant applications. The retreat was the first time people studying the condition from different fields—audiology, neuroscience, and clinical psychology, to name a few—assembled to discuss research priorities. Along with developing and testing treatments, one of those priorities is basic: defining the condition and establishing guidelines for diagnosis.

Whether misophonia should be considered a neurological condition or a psychiatric disorder has been controversial. Because psychiatry deals with problems of behavior and emotion, some psychiatrists see misophonia as within their purview. And although recognition in psychiatry's *Diagnostic and Statistical Manual of Mental Disorders* (DSM) might pave the way for more research and treatment, it could create other problems. Sufferers won't welcome the stigma that might come with a psychiatric diagnosis, however unfair the label might be.

And they've already had a taste of it. In a 2018 episode of the television crime drama *Criminal Minds,* an FBI investigator theorizes that a killer who drills into a victim's auditory cortex could have misophonia: In this case, a violent response to sound. That portrayal did not go over well in the misophonia community, because the condition has thus far not been linked to violence. (As for me, it's difficult enough to be bothered by the sounds *other* people make without also worrying that those same people might fear I'll go berserk over dinner if I hear a trigger noise.)

Negative portrayals aside, it is good to see misophonia getting some attention, especially from scientists. The two studies I've documented here are milestones in determining the condition's cause—but the work continues. I thank any researcher who takes the time to study this mysterious ailment, any therapist who takes it seriously, and any friend or family member who offers support instead of derision. We all have our quirks, and this is mine (or at least, one of mine). I hope discussing it here might spur fellow sufferers to speak up and maybe even reach out for help.

Because let's face it: There's nothing abnormal about being a little abnormal.

AFTERWORD

My favorite scientists are unafraid to talk about the world in blunt terms; they call it like they see it.

I'm thinking here of researchers like Jeff Tomberlin, who makes a passionate and convincing case for maggots to join our food supply. And Helen O'Connell, who was not only willing to map the anatomy of the clitoris, but also to explain to the rest of the medical establishment that, actually, their textbooks are wrong. Or Annalisa Durdle, who is glad to ask volunteers to drop trou and donate some semen, because forensic science needs to feed it to some flies.

This kind of scientific boldness is something the world could use more of. All too often, we're hemmed in by our fears and our sense of propriety. Some subjects simply aren't discussed, or aren't taken seriously as topics of investigation. Sometimes that's because we collectively feel they're gross or embarrassing—like secretions. Or sex. Or dead bodies.

I once had the opportunity to meet the best-selling author Mary Roach—hands down, one of my all-time favorite science writers—who has written with sensitivity and humor about many of these subjects. We commiserated a bit about what it's like to write about "gross stuff": Often, people assume I write for young children, and are forever suggesting book concepts involving farts. But writing about these topics isn't meant merely to make a gross observation and then run away giggling. Rather, the goal is to explore a bigger idea or uncover a fascinating bit of science, to understand ourselves as part of nature. And let's face it: Nature is often gross.

Of course, people don't always enjoy being reminded that they're gross. But we are. We're living, breathing, mucus-producing, defecating, bleeding, and ultimately dying and decomposing animals, and that's nothing to be embarrassed about. We aren't embarrassed about it when we're kids, which is why we associate gross science with children. But as we grow up, we're expected to become ashamed of our bodies, and to lose our curiosity about all things icky.

But I'd say we miss out when we stop being curious about the parts of life that are gross, awkward, or scary. Plenty of scientific breakthroughs began with someone who decided to take a closer look at something others might have turned away from. We wouldn't have antibiotics, for instance, if Alexander Fleming had been too squeamish to collect snot from a patient with a cold. (He noticed the mucus had the odd property of inhibiting bacterial growth.)[*] Later, he was reminded of this booger effect when some mold in one of his petri dishes behaved similarly. That moldy plate was the genesis for his discovery of penicillin, which at first he called "mold juice" (an admirably blunt name, but a hard sell to the public).

Along the same lines, cells from the cervix of a woman named Henrietta Lacks changed medical research forever.[†] These cells collected from a woman's "unmentionables" were the first to be deemed what scientists call immortal, meaning they could divide ad infinitum in the lab without dying. Lacks's cells opened up a new world of cellular experimentation, including tests of the first polio vaccine, the first cellular cloning, and decades of research on cancer and viruses.

....

[*] This was how he discovered lysozyme, an enzyme that occurs naturally in our secretions, including saliva and tears. It helps inhibit bacterial infections in the body, but turned out to be ineffective as an antibiotic drug against disease.

[†] Henrietta Lacks was an African-American woman being treated for cervical cancer at Johns Hopkins University. Some of those cells were collected in 1951 and given to George Otto Gey, a researcher who noticed that they could divide multiple times without dying, which was unusual. He named them HeLa, using the beginning of Lacks's first and last names, and developed a cell line that survives to this day. The cells were taken without Lacks's knowledge, and her family was not informed of the cell line's existence until 1975.

Afterword

Then there's the gross animal world, an endless source of delightful inspiration for engineers and the rest of us. Right now, researchers are working on not only robots that can skitter and compress like roaches, but also bulletproof vests modeled on hagfish slime. They've even figured out how wombats' intestines manufacture their cube-shaped poop.* I can't wait to see what the processed-food companies do with that one, but I'm guessing this could open up a whole new world of extrusion technology—and ultimately, stackable snacks.

I love that animals are gross. It's a testament to life's problem-solving abilities that a hagfish can instantaneously shoot slime capable of ensnaring a shark, and that wombats can mark their territory with a stack of poop that won't roll off a rock. Evolution may not have much concern for dignity, but it sure gets the job done. (I also love that someone had the curiosity to open up a roadkill wombat and figure out how its guts worked.)

This kind of curiosity isn't useful only for the scientists who stumble across dead wombats, and it is not limited to researchers. We could all stand to embrace the grossness in our lives. Think how much easier it would be to survive having a six-year-old in your house if you were comfortable discussing poop. And how nice it would be to see a cockroach and merely think it interesting how it runs just as fast on four legs as six. Not to mention how much easier a conversation with your doctor could be about any number of secretions and bodily functions. I dare say that if we could be more thoughtful about what disgusts us and why, we might even learn to be a little more tolerant of one another. And it all starts with curiosity.

My hope is that reading these stories will reassure you that it's OK to follow your curiosity, even if it takes you briefly into the shadows. As you explore dark places, science can be your

---

* It turns out that a wombat's intestines have areas of alternating stiffness. So as feces travels through, it passes sections that are soft, then stiff, and so on. The stiff areas form the flat faces of the cube.

flashlight, because after all, science is just a way of learning how the world works. And once you know how something works, it's almost always less intimidating.

I've discovered that for myself. I'm no longer terrified to see a dead body, and it got a lot easier to eat an insect when I realized the experience is not gross for everyone. It's also becoming more comfortable to inhabit my female body as I learn to understand it better.

So let's ask more gross questions. Let's be open to discovering the world as it is, not as we wish it were. Let's see the wonder of nature, even if it has six legs or eats its own young. Let's be less afraid, and less ashamed of our bodies. Let's talk about death more.

And most of all, let's be more blunt.

ACKNOWLEDGMENTS

This book would not have been born without my brilliant editor, Hilary Black, who embraced my vision of bringing gross science to a wider audience and expertly guided it into being. As a reward for her efforts, she had to endure endless conversations about bodily fluids and cadavers, and has done so with humor and grace. (I know she likes something when I get an enthusiastic cry of "Vile!") Likewise, much gratitude goes to Michelle Cassidy, who commented on multiple drafts of this book, and Anne Staub, who deftly wrestled much of the text into shape. I feel lucky to have worked with the incredibly talented Briony Morrow-Cribbs, whose extraordinary original artwork perfectly captures the beauty of the beastly. Finally, thanks to the copy editors who combed, cleaned, and conditioned this hairy beast.

Such a bizarre set of interests as mine would never have found an outlet without an incredibly supportive, and relentlessly curious, crew of colleagues. I have to start with my science journalism family at *Science News,* where Gory Details began. Editors Tom Siegfried and Eva Emerson said yes to my idea for a blog about the "dark side of science," perhaps against their better judgment, and for that I will always be grateful. Kate Travis was my indefatigable editor and cheerleader when I needed it most, and helped me think through many a harebrained idea. Reporters Tina Saey, Susan Milius, Bethany Brookshire, Laura Sanders, and many more have all generously shared story ideas, and they weren't even all about poop.

At National Geographic, thanks go to Jamie Shreeve and Dan Gilgoff for hiring me as science editor in part because of this crazy thing I did on the side, instead of running screaming as normal bosses would have done. Thank you to my phenomenal Phenomena blog network colleagues there as well: Nadia Drake, Robert Krulwich, Ed Yong, Carl Zimmer, Maryn McKenna, Virginia Hughes, and Brian Switek. You have all inspired me. Victoria Jaggard is my blog editor extraordinaire and soul mate in morbid curiosity, and many other brilliant colleagues there have fueled me with their smart ideas and unflagging support: Mark Strauss, Michael Greshko, Christine Dell'Amore, Brian Howard, Becky Little, and Laura Parker, to name a few.

One of the perks of being a science journalist is that when researchers do something interesting, a reporter has an excuse to call them up and ask questions that would normally be considered nosy to the point of impoliteness. How on Earth did you come up with this idea? Tell me about the time when [whatever gross thing we're talking about] happened. What did it feel like? And amazingly, they usually tell you. Thank you to all of the scientists who have endured my questions and follow-up questions. There would be nothing of interest for me to write about without your hard work and scientific curiosity. And an extraspecial thank you to those who have invited me into their labs and offices or welcomed me at their gatherings, including Megan Thoemmes and Rob Dunn, Jeff Tomberlin, Nancy Hinkle, Gale Ridge, Bruce Goldfarb and Baltimore's Office of the Chief Medical Examiner, the organizers of the Eating Insects meeting, and the American Academy of Forensic Sciences. Likewise, thank you to the two brave women who spoke with me about their personal experiences with delusory parasitosis. I hope that by telling your stories, we might encourage someone to get the help they need.

Finally, my friends and family are an incredible support system. Thank you to Lynn Addison for serving as both eagle-eyed reader and de facto therapist. Every writer should be so lucky as to have a friend who'll put a glass of wine in her hand and say it will all be OK, and then help hash out whatever writing obstacle has arisen. My husband, Jay, is the great steadying force that allows me to plunge forward without tipping over. And I owe great thanks to my parents. Thank you, Mom, for teaching me to read before the teachers wanted you to, and then telling the librarians to let me check out more than the maximum number of books because you were certain that I would read them all. I always did, and I owe my lifelong love of reading to you. And thank you, Dad, for introducing me to science and believing that I could do it. Your belief that your little girl could understand calculus in grade school, although mistaken, was nevertheless a much appreciated vote of confidence.

SOURCES

PART 1: MORBID CURIOSITY

Introduction: Not Quite CSI

Oosterwijk, S. "Choosing the Negative: A Behavioral Demonstration of Morbid Curiosity." *PLOS One* 12, no. 7 (2017). doi: 10.1371/journal.pone.0178399.

Wilson, Eric G. *Everyone Loves a Good Train Wreck: Why We Can't Look Away.* Sarah Crichton Books, 2012.

The World's Smallest Crime Scenes

Portions originally published as Engelhaupt, Erika. "Peek Into Tiny Crime Scenes Hand-Built by an Obsessed Millionaire." Gory Details (blog), National Geographic, June 15, 2016. https://www.nationalgeographic.com/science/phenomena/2016/06/15/peek-into-tiny-crime-scenes-hand-built-by-an-obsessed-millionaire/

Botz, Corinne. *The Nutshell Studies of Unexplained Death.* The Monacelli Press, 2004.

Goldfarb, Bruce. *18 Tiny Deaths: The Untold Story of Frances Glessner Lee and the Invention of Modern Forensics.* Sourcebooks, 2020.

Detailed images of the Nutshells are available through the Smithsonian Institution's website at *americanart.si.edu/exhibitions/nutshells*.

The Living Dead

Portions originally published as Engelhaupt, Erika. "Getting to Know the Real Living Dead." Used with permission, Gory Details (blog), *Science News*, Nov. 7, 2013. www.sciencenews.org/blog/gory-details/getting-know-real-living-dead; and Engelhaupt, Erika. "You're Surrounded by Bacteria That Are Waiting for You to Die." Gory Details (blog), National Geographic, December 12, 2015. www.nationalgeographic.com/science/phenomena/2015/12/12/youre-surrounded-by-bacteria-that-are-waiting-for-you-to-die.

Bilheux, Hassina Z., et al. "A Novel Approach to Determine Post Mortem Interval Using Neutron Radiography." *Forensic Science International* 251 (June 2015): 11–21.

Costandi, Moheb. "This Is What Happens After You Die." *Mosaic*, May 4, 2015.

Hyde, Embriette R., et al. "The Living Dead: Bacterial Community Structure of a Cadaver at the Onset and End of the Bloat Stage of Decomposition." *PLOS One* 8, no. 10 (October 30, 2013).

Javan, Gulnaz T., et al. "Cadaver Thanatomicrobiome Signatures: The Ubiquitous Nature of Species in Human Decomposition." *Frontiers in Microbiology* 8 (2017): 2096.

Metcalf, Jessica L., et al. "A Microbial Clock Provides an Accurate Estimate of the Postmortem Interval in a Mouse Model System." *Elife* 2 (October 15, 2013).

Pechal, J. L., et al. "Microbial Community Functional Change During Vertebrate Carrion Decomposition." *PLOS One* 8, no 11 (2013). doi: 10.1371/journal.pone.0079035.

Sender, R., et al. "Revised Estimates for the Number of Human and Bacteria Cells in the Body." *PLOS Biology* 14, no. 8 (2016). doi: 10.1371/journal.pbio.1002533.

Vass, Arpad A. "Beyond the Grave—Understanding Human Decomposition." *Microbiology Today* 28 (November 2001): 190–93.

Would Your Dog Eat You If You Died?

Portions originally published as Engelhaupt, Erika. "Would Your Dog Eat You if You Died? Get the Facts." Gory Details (blog), National Geographic, June 23, 2017. news.nationalgeographic.com/2017/06/pets-dogs-cats-eat-dead-owners-forensics-science.

Biro, Dora, et al. "Chimpanzee Mothers at Bossou, Guinea Carry the Mummified Remains of Their Dead Infants." *Current Biology* 20, no. 8 (April 27, 2010): R351–R52.

Buschmann, C., et al. "Post-Mortem Decapitation by Domestic Dogs: Three Case Reports and Review of the Literature." *Forensic Science, Medicine, and Pathology* 7, no. 4 (December 1, 2011): 344–49.

Colard, Thomas, et al. "Specific Patterns of Canine Scavenging in Indoor Settings." *Journal of Forensic Sciences* 60, no. 2 (2015): 495–500.

Sources

Hernández-Carrasco, Mónica, et al. "Indoor Postmortem Mutilation by Dogs: Confusion, Contradictions, and Needs from the Perspective of the Forensic Veterinarian Medicine." *Journal of Veterinary Behavior* 15 (September 1, 2016): 56–60.

King, B. J. *How Animals Grieve*. University of Chicago Press, 2013.

Maksymowicz, Krzysztof, et al. "Refutation of the Stereotype of a 'Killer Dog' in Light of the Behavioral Interpretation of Human Corpses Biting by Domestic Dogs." *Journal of Veterinary Behavior* 6, no. 1 (2011): 50–56.

Ropohl, Dirk, Richard Scheithauer, and Stefan Pollak. "Postmortem Injuries Inflicted by Domestic Golden Hamster: Morphological Aspects and Evidence by DNA Typing." *Forensic Science International* 72, no. 2 (1995): 81–90.

Rothschild, Markus A., and Volkmar Schneider. "On the Temporal Onset of Postmortem Animal Scavenging: 'Motivation' of the Animal." *Forensic Science International* 89, no. 1 (September 19, 1997): 57–64.

Smith, B., ed. *The Dingo Debate: Origins, Behaviour and Conservation*. CSIRO Publishing, 2015.

Steadman, Dawnie Wolfe, and Heather Worne. "Canine Scavenging of Human Remains in an Indoor Setting." *Forensic Science International* 173, no. 1 (2007): 78–82.

Verzeletti, Andrea, Venusia Cortellini, and Marzia Vassalini. "Post-Mortem Injuries by a Dog: A Case Report." *Journal of Forensic and Legal Medicine* 17, no. 4 (2010): 216–19.

The Corpse That Bled

Portions originally published as Engelhaupt, Erika. "How 'Talking' Corpses Were Once Used to Solve Murders." Gory Details (blog), National Geographic, October 9, 2017. news.nationalgeographic.com/2017/10/how-talking-corpses-solve-murders-cruentation-ordeal-science.

Brittain, R. P. "Cruentation in Legal Medicine and in Literature." *Medical History* 9 (1965): 82–88.

James, King I. *Daemonologie, in Forme of a Dialogue*. Printed by Robert Walde-graue, printer to the Kings Majestie, 1597.

Lea, Henry Charles. *Superstition and Force: Essays on the Wager*

of Law-the Wager of Battle-the Ordeal-Torture. 3rd ed. Collins,
1878.

Maeder, Evelyn M., and Richard Corbett. "Beyond Frequency: Perceived Realism and the CSI Effect." *Canadian Journal of Crimi-
nology and Criminal Justice* 57, no. 1 (January 2015): 83–114.

If the Shoe Floats . . .

Anderson, Gail. "Determination of Elapsed Time Since Death in
Homicide Victims Disposed of in the Ocean." Canadian Police
Research Center, 2008.

Anderson, Gail S., and Lynne S. Bell. "Impact of Marine Submer-
gence and Season on Faunal Colonization and Decomposition
of Pig Carcasses in the Salish Sea." *PLOS One* 11, no. 3 (March
1, 2016).

Donoghue, E. R., and S. C. Minnigerode. "Human-Body Buoyancy:
Study of 98 Men." *Journal of Forensic Sciences* 22, no. 3 (1977):
573–79.

Lunetta, Philippe, Curtis Ebbesmeyer, and Jaap Molenaar.
"Behaviour of Dead Bodies in Water." In *Drowning: Prevention,
Rescue, Treatment*, edited by J. Bierens. Springer, 2014, 1149.

PART 2: THAT'S DISGUSTING
Introduction: A Buggy Buffet

Curtis, Valerie. *Don't Look, Don't Touch, Don't Eat.* University of
Chicago Press, 2013.

Ruby, M. B., P. Rozin, and C. Chan. "Determinants of Willingness to
Eat Insects in the USA and India." *Journal of Insects as Food and
Feed* 1, no. 3 (2015): 215–25.

On the Maggot Farm

Evans, Josh, et al. *On Eating Insects: Essays, Stories and Recipes.*
Phaidon Press, 2017.

Huis, Arnold van, et al. "Edible Insects: Future Prospects for Food
and Feed Security." Rome: Food and Agriculture Organization of
the United Nations, 2013.

Sheppard, D. C., J. K. Tomberlin, J. A. Joyce, B. C. Kiser, and S. M.

Sources

Sumner. "Rearing and Colony Maintenance of the Black Soldier Fly, *Hermetia illucens* (L.) (Diptera: Stratiomyidae)." *Journal of Medical Entomology* 39 (2002): 695–98.

Sheppard, C., J. K. Tomberlin, and G. L. Newton. "Use of Soldier Fly Larvae to Reduce Manure, Control House Fly Larvae, and Produce High Quality Feedstuff." Paper presented at the National Poultry Waste Management Symposium, 1998.

Stinks So Good

Portions originally published as Engelhaupt, Erika. "People Sometimes Like Stinky Things—Here's Why." Gory Details (blog), National Geographic, August 3, 2015. www.national geographic.com/science/phenomena/2015/08/03/why-do-people -sometimes-like-stinky-things.

Rozin, Paul, et al. "Glad to Be Sad, and Other Examples of Benign Masochism." *Judgment and Decision Making* 8, no. 4 (July 2013): 439–47.

Toffolo, Marieke B. J., Monique A. M. Smeets, and Marcel A. van den Hout. "Proust Revisited: Odours as Triggers of Aversive Memories." *Cognition & Emotion* 26, no. 1 (2012): 83–92.

Pass the Semen

Portions originally published as Engelhaupt, Erika. "Flies Could Falsely Place Someone at a Crime Scene." Gory Details (blog), National Geographic, February 22, 2016. www.nationalgeographic.com/ science/phenomena/2016/02/22/flies-could-falsely-place-someone -at-a-crime-scene.

Cale, Cynthia. "Forensic DNA Evidence Is Not Infallible." *Nature* 526 (October 29, 2015).

Cale, Cynthia M., et al. "Could Secondary DNA Transfer Falsely Place Someone at the Scene of a Crime?" *Journal of Forensic Sciences* 61, no. 1 (January 2016): 196-203.

Cale, C., et al. "Indirect DNA Transfer: The Impact of Contact Length on Skin-to-Skin-to-Object DNA Transfer." In American Academy of Forensic Sciences annual meeting, Baltimore, 2019.

Durdle, Annalisa, et al. "Location of Artifacts Deposited by the Blow Fly *Lucilia cuprina* After Feeding on Human Blood at Simulated

Indoor Crime Scenes." *Journal of Forensic Sciences* 63, no. 4 (July 2018): 1261–68.

Durdle, Annalisa, Robert J. Mitchell, and Roland A. H. van Oorschot. "The Food Preferences of the Blow Fly *Lucilia cuprina* Offered Human Blood, Semen and Saliva, and Various Nonhuman Foods Sources." *Journal of Forensic Sciences* 61, no. 1 (January 2016): 99–103.

Hussain, Ashiq, et al. "Ionotropic Chemosensory Receptors Mediate the Taste and Smell of Polyamines." *PLOS Biology* 14, no. 5 (May 2016).

Sniffing Out Sickness

Originally published as Engelhaupt, Erika. "You Can Smell When Someone's Sick—Here's How." Gory Details (blog), National Geographic, January 18, 2018. news.nationalgeographic.com/2018/01/smell-sickness-parkinsons-disease-health-science.

Gordon, S. G., et al. "Studies of Trans-3-Methyl-2-Hexenoic Acid in Normal and Schizophrenic Humans." *Journal of Lipid Research* 14, no. 4 (1973): 495–503.

McGann, John P. "Poor Human Olfaction Is a 19th-Century Myth." *Science* 356, no. 6338 (May 12, 2017).

Regenbogen, Christina, et al. "Behavioral and Neural Correlates to Multisensory Detection of Sick Humans." *Proceedings of the National Academy of Sciences of the United States of America* 114, no. 24 (June 13, 2017): 6400–405.

Smith, K., G. F. Thompson, and H. D. Koster. "Sweat in Schizophrenic Patients: Identification of Odorous Substance." *Science* 166, no. 3903 (1969): 398–99.

Trivedi, Drupad K., et al. "Discovery of Volatile Biomarkers of Parkinson's Disease from Sebum." *ACS Central Science* 5, no. 4 (April 24, 2019): 599–606.

Sewer Monsters

Originally published as Engelhaupt, Erika. "Huge Blobs of Fat and Trash Are Filling the World's Sewers." Gory Details (blog), National Geographic, August 16, 2017. news.nationalgeographic.com/2017/08/fatbergs-fat-cities-sewers-wet-wipes-science.

He, Xia, et al. "Evidence for Fat, Oil, and Grease (FOG) Deposit Formation Mechanisms in Sewer Lines." *Environmental Science & Technology* 45, no. 10 (May 15, 2011): 4385–91.

He, Xia, et al. "Mechanisms of Fat, Oil and Grease (FOG) Deposit Formation in Sewer Lines." *Water Research* 47, no. 13 (September 1, 2013): 4451–59.

PART 3: BREAKING TABOOS

Introduction: The Ultimate Taboos

Beaglehole, J. C., ed. *The Journals of Captain James Cook on His Voyages of Discovery*. Vol. 1: Hakluyt Society, 2017.

Bergmann, Anna. "Taboo Transgressions in Transplantation Medicine." *Journal of American Physicians and Surgeons* 13, no. 2 (2008): 52–55.

Canavero, Sergio. "The 'Gemini' Spinal Cord Fusion Protocol: Reloaded." *Surgical Neurology International* 6 (2015): 18.

———. "Heaven: The Head Anastomosis Venture Project Outline for the First Human Head Transplantation with Spinal Linkage (Gemini)." *Surgical Neurology International* 4, Suppl. 1 (2013): S335–42.

Niu, A., et al. "Heterotopic Graft of Infant Rat Brain as an Ischemic Model for Prolonged Whole-Brain Ischemia." *Neuroscience Letters* 325, no. 1 (May 31, 2002): 37–41.

Ren, Xiao-Ping, et al. "Allogeneic Head and Body Reconstruction: Mouse Model." *CNS Neuroscience & Therapeutics* 20, no. 12 (December 2014): 1056–60.

It's Hard to Get a Head

Portions originally published as "Surgeon Reveals Head Transplant Plan, But Patient Steals the Show." Gory Details (blog), National Geographic, June 12, 2015. www.nationalgeographic.com/science/phenomena/2015/06/12/surgeon-reveals-head-transplant-plan-but-patient-steals-the-show; and Engelhaupt, Erika. "Human Head Transplant Proposed—How Did We Get Here?" Gory Details (blog), National Geographic, May 5, 2015. www.nationalgeographic.com/science/phenomena/2015/05/05/human-head-transplant-proposed-how-did-we-get-here.

Konstantinov, I. E. "At the Cutting Edge of the Impossible: A Tribute to Vladimir P. Demikhov." *Texas Heart Institute Journal* 36, no. 5 (October 2009): 453–58.

Ren, Xiaoping, and Sergio Canavero. "Heaven in the Making: Between the Rock (the Academe) and a Hard Case (a Head Transplant)." *AJOB Neuroscience* 8, no. 4 (October 2, 2017): 200–205.

White, R. J., et al. "Brain Transplantation: Prolonged Survival of Brain after Carotid-Jugular Interposition." *Science* 150, no. 3697 (1965): 779–81.

White, R. J., et al. "Cephalic Exchange Transplantation in the Monkey." *Surgery* 70, no. 1 (1971): 135–39.

The Most Murderous Mammals

Portions originally published as Engelhaupt, Erika. "How Human Violence Stacks Up Against Other Killer Animals." Gory Details (blog), National Geographic, September 28, 2016. news.national geographic.com/2016/09/human-violence-evolution-animals -nature-science.

Bell, M. B. V., et al. "Suppressing Subordinate Reproduction Provides Benefits to Dominants in Cooperative Societies of Meerkats." *Nature Communications* 5 (July 2014).

Gomez, J. M., et al. "The Phylogenetic Roots of Human Lethal Violence." *Nature* 538, no. 7624 (October 2016): 233–37.

Perrtree, R. M., et al. "First Observed Wild Birth and Acoustic Record of a Possible Infanticide Attempt on a Common Bottlenose Dolphin *(Tursiops truncatus)*." *Marine Mammal Science* 32, no. 1 (January 2016): 376–85.

United Nations Office on Drugs and Crime. *Global Study on Homicide,* 2013.

Wrangham, Richard. *The Goodness Paradox.* Knopf Doubleday, 2019.

A Practical Guide to Cannibalism

Portions originally published as Engelhaupt, Erika. "Some Animals Eat Their Moms, and Other Cannibalism Facts." Used with permission, Gory Details (blog), *Science News,* February 6, 2014. www.sciencenews.org/blog/gory-details/some-animals-eat -their-moms-and-other-cannibalism-facts; and Engelhaupt,

Erika. "Cannibalism Study Finds People Are Not That Nutritious": Gory Details (blog), National Geographic, April 6, 2017. news.national geographic.com/2017/04/human-cannibalism -nutrition-archaeology-science.

Carbonell, E., et al. "Cultural Cannibalism as a Paleoeconomic System in the European Lower Pleistocene." *Current Anthropology* 51, no. 4 (August 2010): 539–49.

Cole, J. "Assessing the Calorific Significance of Episodes of Human Cannibalism in the Palaeolithic." *Scientific Reports* 7 (April 2017).

Evans, T. A., E. J. Wallis, and M. A. Elgar. "Making a Meal of Mother." *Nature* 376, no. 6538 (July 1995): 299.

Schutt, Bill. *Cannibalism: A Perfectly Natural History.* Algonquin Books, 2017.

Soulsby, James. *Animal Cannibalism: The Dark Side of Evolution.* 5m Publishing, 2013.

Pioneering the Clitoris

Blechner, Mark J. "The Clitoris: Anatomical and Psychological Issues." *Studies in Gender and Sexuality* 18, no. 3 (2017): 190–200.

Buisson, Odile, et al. "Coitus as Revealed by Ultrasound in One Volunteer Couple." *The Journal of Sexual Medicine* 7, no. 8 (2010): 2750–54.

Di Marino, Vincent, and Hubert Lepidi. *Anatomic Study of the Clitoris and the Bulbo-Clitoral Organ,* Vol. 91. Springer, 2014.

Ehrenreich, Barbara, and Deirdre English. *Complaints and Disorders: The Sexual Politics of Sickness.* 2nd ed. The Feminist Press at CUNY, 2011.

Fyfe, Melissa. "Get Cliterate: How a Melbourne Doctor Is Redefining Female Sexuality." The *Sydney Morning Herald,* December 8, 2018.

Hoag, Nathan, Janet R. Keast, and Helen E. O'Connell. "The 'G-Spot' Is Not a Structure Evident on Macroscopic Anatomic Dissection of the Vaginal Wall." *The Journal of Sexual Medicine* 14, no. 12 (2017): 1524–32.

Moore, Lisa Jean, and Adele E. Clarke. "Clitoral Conventions and Transgressions: Graphic Representations in Anatomy Texts, c1900-1991." *Feminist Studies* 2, (Summer 1995): 255–301.

Park, Katharine. "The Rediscovery of the Clitoris: French Medicine and the Tribade." In *The Body in Parts: Fantasies of Corporeality in Early Modern Europe,* edited by Carla Mazzio and David Hillman. Routledge, 1997, 171–93.

Pfaus, James G., et al. "The Whole Versus the Sum of Some of the Parts: Toward Resolving the Apparent Controversy of Clitoral Versus Vaginal Orgasms." *Socioaffective Neuroscience & Psychology* 6, no. 1 (2016): 32578.

Sheehan, Elizabeth. "Victorian Clitoridectomy: Isaac Baker Brown and His Harmless Operative Procedure." *Medical Anthropology Newsletter* 12, no. 4 (1981): 9–15.

Stringer, Mark D., and Ines Becker. "Colombo and the Clitoris." *European Journal of Obstetrics & Gynecology and Reproductive Biology* 151, no. 2 (2010): 130-33.

Volck, William, et al. "Gynecologic Knowledge Is Low in College Men and Women." *Journal of Pediatric and Adolescent Gynecology* 26, no. 3 (2013): 161–66.

One Giant Leap for Womankind

Portions originally published as Engelhaupt, Erika. "How Do Women Deal With Having a Period . . . in Space?" Gory Details (blog), National Geographic, April 22, 2016. www.nationalgeographic.com/science/phenomena/2016/04/22/how-do-women-deal-with-having-a-period-in-space.

Jain, Varsha, and Virginia E. Wotring. "Medically Induced Amenorrhea in Female Astronauts." *NPJ Microgravity* 2 (2016): article no. 16008.

Ride, Sally K. "NASA Johnson Space Center Oral History Project, Edited Oral History Transcript." Interviewed by Rebecca Wright, October 22, 2002. historycollection.jsc.nasa.gov/JSCHistoryPortal/history/oral_histories/RideSK/RideSK_10-22-02.htm.

Ne-crow-philia

Harrison, Ben. *Undying Love: The True Story of a Passion That Defied Death.* Macmillan, 2001.

"Hold Von Cosel on Malicious and Wanton Charges." *The Key West Citizen*, October 7, 1940.

Langley, Liz. *Crazy Little Thing: Why Love and Sex Drive Us Mad.* Cleis Press, 2011.

Moeliker, C. W. "The First Case of Homosexual Necrophilia in the Mallard *Anas platyrhynchos* (Aves: Anatidae)." *Deinsea* 8, no. 1 (2001): 243–48.

Rosman, Jonathan P., and Phillip J. Resnick. "Sexual Attraction to Corpses: A Psychiatric Review of Necrophilia." *Bulletin of the American Academy of Psychiatry and the Law* 17, no. 2 (1989): 153–63.

Russell, Douglas G. D., William J. L. Sladen, and David G. Ainley. "Dr. George Murray Levick (1876–1956): Unpublished Notes on the Sexual Habits of the Adélie Penguin." *Polar Record* 48, no. 4 (2012): 387–93.

Troyer, John. "Abuse of a Corpse: A Brief History and Re-Theorization of Necrophilia Laws in the USA." *Mortality* 13, no. 2 (2008): 132–52.

PART 4: CREEPY CRAWLIES

Introduction: Bug Off!

Portions originally published as Engelhaupt, Erika. "This Is What Happens When You Use Rat Poison: Flymageddon." Gory Details (blog), National Geographic, May 15, 2015. www.nationalgeographic.com/science/phenomena/2015/05/15/this-is-what-happens-when-you-use-rat-poison-flymageddon.

Bass, Bill, William M. Bass, and Jon Jefferson. *Death's Acre: Inside the Legendary Forensic Lab—the Body Farm—Where the Dead Do Tell Tales.* Penguin, 2004.

Rat Race

Originally published as Engelhaupt, Erika. "Yes, Rats Can Swim Up Your Toilet. And It Gets Worse Than That." Gory Details (blog), National Geographic, August 14, 2015. www.nationalgeographic.com/science/phenomena/2015/08/14/yes-rats-can-swim-up-your-toilet-and-it-gets-worse-than-that.

Small, but Mitey

Dunn, Robert R. "The Evolution of Human Skin and the Thousands of Species It Sustains, With Ten Hypothesis of Relevance to Doctors." In *Personalized, Evolutionary, and Ecological Dermatology*, edited by R. A. Norman. Switzerland: Springer International Publishing, 2016.

Moran, Ellen M., Ruth Foley, and Frank C. Powell. "*Demodex* and Rosacea Revisited." *Clinics in Dermatology* 35, no. 2 (2017): 195–200.

Palopoli, Michael F., et al. "Global Divergence of the Human Follicle Mite *Demodex folliculorum:* Persistent Associations Between Host Ancestry and Mite Lineages." *Proceedings of the National Academy of Sciences* 112, no. 52 (2015): 15958–63.

Thoemmes, Megan S., et al. "Ubiquity and Diversity of Human-Associated *Demodex* Mites." *PLOS One* 9, no. 8 (2014): e106265.

Roaches, Debunked

Portions originally published as Engelhaupt, Erika. "Amazing Video Reveals Why Roaches Are So Hard to Squish." Gory Details (blog), National Geographic, February 8, 2016. www.national geographic.com/science/phenomena/2016/02/08/watch-amazing -video-reveals-why-roaches-are-so-hard-to-squish.

Fardisi, Mahsa, et al. "Rapid Evolutionary Responses to Insecticide Resistance Management Interventions by the German Cockroach (*Blattella germanica* L.)." *Scientific Reports* 9, no. 1 (2019): 8292.

Jayaram, Kaushik, et al. "Transition by Head-on Collision: Mechanically Mediated Manoeuvres in Cockroaches and Small Robots." *Journal of The Royal Society Interface* 15, no. 139 (2018): 20170664.

Jayaram, Kaushik, and Robert J. Full. "Cockroaches Traverse Crevices, Crawl Rapidly in Confined Spaces, and Inspire a Soft, Legged Robot." *Proceedings of the National Academy of Sciences* 113, no. 8 (2016): E950–E957.

Jindrich, Devin L., and Robert J. Full. "Dynamic Stabilization of Rapid Hexapedal Locomotion." *Journal of Experimental Biology* 205, no. 18 (2002): 2803–23.

Schweid, Richard. *The Cockroach Papers: A Compendium of History and Lore.* Four Walls Eight Windows, 1999.

Sources

I've Got You Under My Skin

Portions originally published as Engelhaupt, Erika. "A Horrifying List of Creatures That Can Crawl Into Your Body." Gory Details (blog), National Geographic, February 14, 2017. news.nationalgeographic .com/2017/02/roach-in-nose-ear-insects-science.

Morris, Thomas. "Worms in the Nose." *Thomas-Morris.uk*, 2016.

O'Toole, K., et al. "Removing Cockroaches From the Auditory Canal: Controlled Trial." *The New England Journal of Medicine* 312, no. 18 (1985): 1197–97.

That's Not an Eyelash

Bradbury, Richard S., et al. "Case Report: Conjunctival Infestation with Thelazia Gulosa: A Novel Agent of Human Thelaziasis in the United States." *American Journal of Tropical Medicine and Hygiene* 98, no. 4 (2018): 1171–74.

Worms on the Brain

Originally published as Engelhaupt, Erika. "Parasitic Worms Found in a Woman's Eye—First Case of Its Kind." Gory Details (blog), National Geographic, February 12, 2018. news.nationalgeographic.com/2018/ 02/eye-worms-parasites-oregon-thelazia-gulosa-health-science.

Thiengo, Silvana Carvalho, et al. "*Angiostrongylus cantonensis* and Rat Lungworm Disease in Brazil." *Hawai'i Journal of Medicine & Public Health* 72, no. 6, Suppl. 2 (2013): 18.

Wang, Huijie, et al. "Eating Centipedes Can Result in *Angiostrongylus cantonensis* Infection: Two Case Reports and Pathogen Investigation." *The American Journal of Tropical Medicine and Hygiene* 99, no. 3 (2018): 743–48.

The World's Worst Sting

Originally published as Engelhaupt, Erika. "This Is the Worst Insect Sting in the World." Gory Details (blog), National Geographic, September 26, 2016. news.nationalgeographic.com/2016/09/ worlds-most-painful-insect-sting-science.

Schmidt, Justin O. *The Sting of the Wild.* JHU Press, 2016.

Smith, Michael L. "Honey Bee Sting Pain Index by Body Location." *PeerJ* 2 (2014): e338.

Wilcox, Christie. *Venomous: How Earth's Deadliest Creatures Mastered Biochemistry*. Scientific American/Farrar, Straus and Giroux, 2016.

PART 5: GROSS ANATOMY

Introduction: To Secrete or Excrete

Yang, Patricia J., et al. "Hydrodynamics of Defecation." *Soft Matter* 13, no. 29 (2017): 4960–70.

Digging for Gold

Burres, Steven. 2011. Device and method for removing earwax. U.S. Patent application US20120296355A1. patents.google.com/patent/US20120296355A1/en?oq=US20120296355A1.

Ernst, E. "Ear Candles: A Triumph of Ignorance over Science." *The Journal of Laryngology & Otology* 118, no. 1 (2004): 1–2.

Hinde, Alfred. "An Efficient and Inexpensive Instrument for the Removal of Ear Wax." *Journal of the American Medical Association* 28, no. 19 (1897): 908–08.

Pahuja, Deepak, and Ryan Scott Bookhamer. 2012. Disposable dual-tipped ear curette incorporating depth measurement system. U.S. Patent US20130190647A1. patents.google.com/patent/US20130190647A1/en.

Prokop-Prigge, Katharine A., et al. "Identification of Volatile Organic Compounds in Human Cerumen." *Journal of Chromatography B* 953 (2014): 48–52.

Prokop-Prigge, Katharine A., et al. "Ethnic/Racial and Genetic Influences on Cerumen Odorant Profiles." *Journal of Chemical Ecology* 41, no. 1 (2015): 67–74.

Shokry, Engy, and Nelson Roberto Antoniosi Filho. "Insights into Cerumen and Application in Diagnostics: Past, Present and Future Prospective." *Biochemia Medica* 27, no. 3 (2017): 1–15.

Yao, Kou C., and Yao Nancy Combined toothbrush, tongue scraper, and ear cleaner. U.S. Patent US3254356A. patents.google.com/patent/US3254356A/en.

The Fecal Cure

Portions originally published as Engelhaupt, Erika. "Introducing the First Bank of Feces." Used with permission, Gory Details (blog), *Science*

News, February 12, 2014. www.sciencenews.org/blog/gory-details/
introducing-first-bank-feces; and as "Alternatives Needed to
Do-It-Yourself Feces Swaps." Used with permission, Gory Details
(blog), *Science News,* February 20, 2014. www.sciencenews.org/
blog/gory-details/alternatives-needed-do-it-your self-feces-swaps.

Falkow, S. "Fecal Transplants in the 'Good Old Days'." Small Things
Considered (blog), 2013.

Jia, N. "A Misleading Reference for Fecal Microbiota Transplant." *The
American Journal of Gastroenterology* 110, no. 12 (2015): 1731.

Juul, Frederik E., et al. "Fecal Microbiota Transplantation for
Primary *Clostridium difficile* Infection." *New England Journal
of Medicine* 378, no. 26 (2018): 2535–36.

Meyers, Shawn, et al. "Clinical Inquiries: How Effective and Safe Is
Fecal Microbial Transplant in Preventing *C. difficile* Recur-
rence?" *The Journal of Family Practice* 67, no. 6 (2018):
386–88.

Saey., T. S. "A Gut Bacteria Transplant May Not Help You Lose
Weight." *Science News,* May 9, 2019.

Zhang, Faming, et al. "Should We Standardize the 1,700-Year-Old
Fecal Microbiota Transplantation?" *The American Journal of
Gastroenterology* 107, no. 11 (2012): 1755.

Pee in the Pool

Portions originally published as Engelhaupt, Erika. "This Is What
Happens When You Pee in the Pool." Used with permission, Gory
Details (blog), *Science News,* April 8, 2014. www.sciencenews
.org/blog/gory-details/what-happens-when-you-pee-pool.

Andersson, Martin, et al. "Early Life Swimming Pool Exposure and
Asthma Onset in Children—a Case-Control Study." *Environmental Health* 17, no. 1 (2018): 34.

Lian, Lushi, et al. "Volatile Disinfection Byproducts Resulting from
Chlorination of Uric Acid: Implications for Swimming Pools."
Environmental Science & Technology 48, no. 6 (2014): 3210–17.

Licking Your Wounds

Originally published as Engelhaupt, Erika. "How Dog and Cat 'Kisses'
Can Turn Deadly." Gory Details (blog), National Geographic,

October 24, 2017. news.nationalgeographic.com/2017/10/
dogs-cats-clean-licking-bacteria-health-science.

Buma, Ryoko, et al. "Pathogenic Bacteria Carried by Companion
Animals and Their Susceptibility to Antibacterial Agents."
Biocontrol Science 11, no. 1 (2006): 1–9.

Butler, T. "*Capnocytophaga canimorsus:* An Emerging Cause of Sep-
sis, Meningitis, and Post-Splenectomy Infection After Dog Bites."
European Journal of Clinical Microbiology & Infectious Diseases
34, no. 7 (2015): 1271–80.

Dewhirst, Floyd E., et al. "The Canine Oral Microbiome." *PLOS One*
7, no. 4 (2012): e36067.

Dewhirst, Floyd E., et al. "The Feline Oral Microbiome: A Provisional 16s
rRNA Gene Based Taxonomy With Full-Length Reference
Sequences." *Veterinary Microbiology* 175, no. 2–4 (2015): 294–303.

Van Dam, A. P., and A. Jansz. "*Capnocytophaga canimorsus* Infections
in the Netherlands: A Nationwide Survey." *Clinical Microbiology
and Infection* 17, no. 2 (2011): 312–15.

To Pee or Not to Pee?

Originally published as Engelhaupt, Erika. "Urine Is Not Sterile, and
Neither Is the Rest of You." Used with permission, Gory Details
(blog), *Science News,* May 22, 2014. www.sciencenews.org/blog/
gory-details/urine-not-sterile-and-neither-rest-you.

Branton, William G., et al. "Brain Microbial Populations in HIV/AIDS:
Alpha-Proteobacteria Predominate Independent of Host Immune
Status." *PLOS One* 8, no. 1 (January 23, 2013).

Branton, W. G., et al. "Brain Microbiota Disruption Within Inflammatory Demyelinating Lesions in Multiple Sclerosis." *Scientific
Reports* 6 (November 28, 2016).

Putnam, David F. "Composition and Concentrative Properties of
Human Urine." NASA, 1971.

Stout, Molly J., et al. "Identification of Intracellular Bacteria in the
Basal Plate of the Human Placenta in Term and Preterm Gesta-
tions." *American Journal of Obstetrics and Gynecology* 208,
no. 3 (2013): 226.e1–26.e7.

Thomas-White, Krystal, et al. "Culturing of Female Bladder Bacteria

Reveals an Interconnected Urogenital Microbiota." *Nature Communications* 9, no. 1557 (2018).

The Need to Bleed

Portions originally published as Engelhaupt, Erika. "Bloodletting Is Still Happening, Despite Centuries of Harm." Gory Details (blog), National Geographic, October 27, 2015. www.nationalgeographic .com/science/phenomena/2015/10/27/bloodletting-is-still-happen ing-despite-centuries-of-harm.

Bennett, J. Hughes. "Further Observations on the Restorative Treatment of Pneumonia." *The Lancet* 87, no. 2214 (1866): 114–16.

Fernandez-Real, J. M., et al. "Blood Letting in High-Ferritin Type 2 Diabetes: Effects on Insulin Sensitivity and Beta-Cell Function." *Diabetes* 51, no. 4 (April 2002): 1000–1004.

Morens, D. M. "Death of a President." *New England Journal of Medicine* 341, no. 24 (December 9, 1999): 1845–50.

North, R. L. "Benjamin Rush, MD: Assassin or Beloved Healer?" *Baylor University Medical Center Proceedings* 13, no. 1 (January 2000): 45–49.

Parapia, Liakat Ali. "History of Bloodletting by Phlebotomy." *British Journal of Haematology* 143, no. 4 (November 2008): 490–95.

Thomas, D. P. "The Demise of Bloodletting." *The Journal of the Royal College of Physicians of Edinburgh* 44, no. 1 (2014): 72–77.

The Detox Myth

Portions originally published as Engelhaupt, Erika. "Fact or Fiction: Can You Really Sweat Out Toxins?" Gory Details (blog), National Geographic, April 6, 2018. news.nationalgeographic.com/2018/04/ sweating-toxins-myth-detox-facts-saunas-pollutants-science.

Imbeault, Pascal, Nicholas Ravanelli, and Jonathan Chevrier. "Can POPs Be Substantially Popped out through Sweat?" *Environment International* 111 (February 2018): 131–32.

PART 6: MYSTERIOUS MINDS

Introduction: A Bug in the Program

Lindner, Isabel, et al. "Observation Inflation: Your Actions Become

Mine." *Psychological Science* 21, no. 9 (September 2010): 1291–99.

The Invisibugs

Portions originally published as Engelhaupt, Erika. "Delusions of Infestation Aren't as Rare as You'd Think." Gory Details (blog), National Geographic, June 22, 2018. news.nationalgeographic .com/2018/06/delusions-infestation-insects-skin-ekboms-syndrome -health-science.

Kohorst, John J., et al. "Prevalence of Delusional Infestation—a Population-Based Study." *JAMA Dermatology* 154, no. 5 (May 2018): 615–17.

The Voodoo Doll Riddle

Portions originally published as Engelhaupt, Erika. "Why Stabbing a Voodoo Doll Is So Satisfying." Used with permission, Gory Details (blog), *Science News,* June 5, 2014. www .sciencenews.org/blog/gory-details/why-stabbing-voodoo-doll -so-satisfying.

Ayduk, Oezlem, Anett Gyurak, and Anna Luerssen. "Individual Differences in the Rejection-Aggression Link in the Hot Sauce Paradigm: The Case of Rejection Sensitivity." *Journal of Experimental Social Psychology* 44, no. 3 (May 2008): 775–82.

Bushman, Brad J., et al. "Low Glucose Relates to Greater Aggression in Married Couples." *Proceedings of the National Academy of Sciences of the United States of America* 111, no. 17 (April 29, 2014): 6254–57.

DeWall, C. Nathan, et al. "The Voodoo Doll Task: Introducing and Validating a Novel Method for Studying Aggressive Inclinations." *Aggressive Behavior* 39, no. 6 (November 2013): 419–39.

Huber, R., D. L. Bannasch, and P. Brennan. "Aggression." In *Advances in Genetics* 75 (2011): 2–293.

Liang, Lindie H., et al. "Righting a Wrong: Retaliation on a Voodoo Doll Symbolizing an Abusive Supervisor Restores Justice." *Leadership Quarterly* 29, no. 4 (August 2018): 443–56.

Lieberman, J. D., et al. "A Hot New Way to Measure Aggression: Hot Sauce Allocation." *Aggressive Behavior* 25, no. 5 (1999): 331–48.

Rozin, P., L. Millman, and C. Nemeroff. "Operation of the Laws of Sympathetic Magic in Disgust and Other Domains." *Journal of Personality and Social Psychology* 50, no. 4 (April 1986): 703–12.

Back Off, Bozo

Portions originally published as Engelhaupt, Erika. "The Real Reason Clowns Creep Us Out." Gory Details (blog), National Geographic, October 7, 2016. news.nationalgeographic.com/2016/10/real-reason-clowns-creep-us-out.

McAndrew, Francis T., and Sara S. Koehnke. "On the Nature of Creepiness." *New Ideas in Psychology* 43 (December 2016): 10–15.

Never Forget a Face

Portions originally published as Engelhaupt, Erika. "Do You Have a Face-Finding Superpower for Fighting Crime?" Gory Details (blog), National Geographic, September 1, 2015. www.nationalgeographic.com/science/phenomena/2015/09/01/do-you-have-a-face-finding-superpower-for-fighting-crime.

Elbich, Daniel B., and Suzanne Scherf. "Beyond the FFA: Brain-Behavior Correspondences in Face Recognition Abilities." *Neuroimage* 147 (February 15, 2017): 409–22.

Russell, Richard, Brad Duchaine, and Ken Nakayama. "Super-Recognizers: People with Extraordinary Face Recognition Ability." *Psychonomic Bulletin & Review* 16, no. 2 (April 2009): 252–57.

Sacks, Oliver. "Face-Blind: Why Are Some of Us Terrible at Recognizing Faces?" *New Yorker,* August 23, 2010, 36–43.

White, David, et al. "Perceptual Expertise in Forensic Facial Image Comparison." *Proceedings of the Royal Society B: Biological Sciences* 282, no. 1814 (2015): 20151292.

Celluloid Psychos

Portions originally published as Engelhaupt, Erika. "The Most (and Least) Realistic Movie Psychopaths Ever." Used with permission, Gory Details (blog), *Science News,* January 14, 2014. www.sciencenews.org/blog/gory-details/most-and-least-realistic-movie-psychopaths-ever.

Fallon, James. *The Psychopath Inside: A Neuroscientist's Personal Journey into the Dark Side of the Brain.* Current, 2014.

Hare, Robert D., and Craig S. Neumann. "Psychopathy: Assessment and

Forensic Implications." *Canadian Journal of Psychiatry/Revue Canadienne De Psychiatrie* 54, no. 12 (December 2009): 791–802.

Leistedt, Samuel J., and Paul Linkowski. "Psychopathy and the Cinema: Fact or Fiction?" *Journal of Forensic Sciences* 59, no. 1 (January 2014): 167–74.

Lilienfeld, Scott O., et al. "Correlates of Psychopathic Personality Traits in Everyday Life: Results from a Large Community Survey." *Frontiers in Psychology* 5 (July 22, 2014).

Ogloff, James R. P. "Psychopathy/Antisocial Personality Disorder Conundrum." *Australian and New Zealand Journal of Psychiatry* 40, no. 6–7 (June–July 2006): 519–28.

The Sound and the Fury

Edelstein, Miren, et al. "Misophonia: Physiological Investigations and Case Descriptions." *Frontiers in Human Neuroscience* 7 (June 25, 2013).

Kumar, Sukhbinder, et al. "The Brain Basis for Misophonia." *Current Biology* 27, no. 4 (February 20, 2017): 527–33.

Kumar, Sukhbinder, et al. "Features Versus Feelings: Dissociable Representations of the Acoustic Features and Valence of Aversive Sounds." *Journal of Neuroscience* 32, no. 41 (October 10, 2012): 14184–92.

AFTERWORD

Skloot, Rebecca. *The Immortal Life of Henrietta Lacks.* Broadway Books, 2017.

Yang, P., et al. "How Do Wombats Make Cubed Poo?" Paper presented at the Annual Meeting of the American Physical Society Division of Fluid Dynamics, Atlanta, 2018.

INDEX

Index

identity and 106
taboos 104–107
Trautwein, Michelle
  173–174
Trinkaus, Erik 128
Troyer, John 152–153
Typhoid, smell of 91

U

Underwater decomposition 56–59
Upton, Liz 81
Urine
  drinking of 231–232
  introduction 208
  smell 211–212
  sterility of, as myth 231–235
  in swimming pools 13, 221–224
Uterus 233–234

V

Vancouver Island, British Columbia 55–61
Vancouver Rat Project 165
Vass, Arpad 40–41
Vesalius, Andreas 132–133
Volatile compounds 93–94
Von Cosel, Carl 151–152
Voodoo Doll Task 263–268

W

Wahl, Mike 30–31
Walden, Heather Stockdale 196–198
Wall Street (movie) 284

Wallace, Sophia 138
Wallace, Thomas 98
Wari' tribe 129
Warrior wasps 201
Washington, George 239
Wasp stings 199, 201
Water, human decomposition in 56–59
Water bugs 178–183
Werdnig-Hoffmann disease 106, 109, 114
Whales, violence in 121
White, David 275–278
White, Robert 115–116
Whiting, Nathaniel 98
Williams, Ted 112
Witches, float test 51, 51n
The Wolf of Wall Street (movie) 284
Wombats 301, 301n
Women
  as astronauts 141–143
  clitoris 131–139
  menstruation 107, 141–145
Worms, human body invasions 191–194
Wotring, Virginia 142, 143
Wrangham, Richard 122–123

X

X-rays 21, 21n

Y

Yazedjian, Laura 56–59
Yellow fever 91, 238–239

"Yellow soup" 215
Yoon, Joseph 69

Z

al Zahrawi 132
Zalkalns, Arnis 274